the Inner Life *of* Cats

the
Inner Life *of* Cats

The Science and Secrets of
Our Mysterious Feline Companions

℘

THOMAS McNAMEE

NEW YORK   BOSTON

Hachette Books
Hachette Book Group
1290 Avenue of the Americas
New York, NY 10104
hachettebooks.com
twitter.com/hachettebooks

First Edition: March 2017

Interior photos by Richard Neill/Adventure Pictures

Hachette Books is a division of Hachette Book Group, Inc.
The Hachette Books name and logo are trademarks of
Hachette Book Group, Inc.

The publisher is not responsible for websites (or their content) that are not owned by the publisher.

The Hachette Speakers Bureau provides a wide range of authors for speaking events. To find out more, go to www.hachettespeakersbureau.com or call (866) 376-6591.

Library of Congress Control Number: 2017930389

ISBNs: 978-0-316-26287-3 (hardcover), 978-0-316-26286-6 (ebook)

Printed in the United States of America

LSC-C

10 9 8 7 6 5 4 3 2 1

For Isabel

Contents

Augusta.

Chapter One

The Kitten

The little black kitten had never felt snow. The snow was fresh-fallen powder, and a single pair of tire tracks led from the gate to the ranch buildings. She called for her mother. She tried to climb out of the tire track, but the snow crumbled beneath her paws and she fell back in. She could not walk very well, but she could walk, so she did, crying for her mother with every step. Night was falling.

She came to a building that smelled of food, and there were people inside, but there was also a dog barking. She continued along the tire track and came to a bridge. After a long moment of hesitation, feeling the deep booming of the water below, the kitten hurried forward, silent now. She followed the track to a wooden wall and could follow it no farther. She struggled along the wall through stiff grass and sharp-crusted snow, afraid, calling again for her mother. She found a little opening in the wall, just whisker-wide, and slipped through. The building smelled of strange things but was not so terribly cold.

She found a pile of rags, licked the snow from between her toes, and fell asleep.

The next morning, a still, cold morning, I saw a little black shape darting amid the unidentifiable junk in the equipment barn.

A few clumsy minutes of chase and I had captured the kitten, who quieted as I held her with one hand and tucked her into the front of my jacket. She was shivering.

There was no answer when Elizabeth and I telephoned our only neighbor, an old Norwegian bachelor rancher, the only conceivable source for the miraculous appearance of this tiny furball. I went out looking for tracks, but there weren't any leading from his place. Then I found the kitten's footprints, smaller than a dime, still brightly visible in the snow all the way down our driveway from the county road, more than a quarter of a mile. There were tire tracks in the road showing that a vehicle had turned around. Someone—who? why? in the middle of the night, in a blizzard— had just dumped her there. What kind of person could do such a thing? We were twenty miles of icy dirt road from the nearest town, Livingston, Montana.

She gobbled the tuna and leftover chicken without looking up, and slurped the milk till her belly pooched out tight. It had a little white star at the center, matching the prim white bow tie at her sternum. She was otherwise all black. I cut down a wine box and filled it with spruce needles and leaf duff, and she used it right away, scratching busily to hide her deposit, watching us with wide eyes as we watched her back in proud delight.

We had no call to be so proud, for all cats naturally use litter boxes at first exposure as long as they feel calm and welcome, because the litter box sufficiently resembles the scent-marking spot that all their kind, including their wild ancestors, employ to proclaim, "This is home."

We exiled her from the bedroom that night, but in the morning she greeted us at the refrigerator with her black stump of a tail straight up and vibrating—the universal cat gesture of greeting and confidence. Her evident understanding of indoor life and her readiness to be cradled in palm or elbow made clear that this was no barn cat, nor victim of abuse. The vet in town declared her female, healthy, about three months old, and likely to be "on the energetic side." Her good cheer, he said, was probably partly due to having been tenderly handled as a baby and partly to an inborn sunny nature, a roll of the genetic dice.

Counting three months back made her birthday in August. We gave our gift from the god of chance the name Augusta.

Augusta's first order of business on the ranch was to make a mental map of her new home. It was more than geographic. In the first days after her arrival, the snow kept her confined to the house, and she spent hour on hour tracing the contours of every chair, table, bookcase, telephone, carpet edge, window frame, and pencil holder. She also was also mapping things invisible to people—mouse trails, bug spoor, hundreds of scents tracked in from outside.

All this was largely an exercise in olfactory memory. Only when each scent impression is recorded deep in the hippocampus is it coordinated with an image in the visual region of the cerebral cortex. That coordination is then deeply engrained. The power of your cat's memory is evident when he startles at the slightest change in his customary surroundings. Cats have five times more surface area inside their noses than we do, and three times more receptor nerves per unit of area lining that surface. Of those receptors there are hundreds of types—the possible combinations of scent perceptions, therefore, being nearly infinite.

A cat's hearing is extremely acute—the widest range, in fact, of any mammal except bats. It is equally superior in its directionality: Watch your cat's ears twitch and swivel, each moving independent of the other, as she tunes in on the precise place from which a sound is coming. For Augusta the faintest whisper of breeze on a screen, swirl of snow, or far call of raven was matter for instant, rapt attention. She seemed pure awareness.

During her explorations Augusta did not care to be interrupted.

She knew her name very quickly, but she responded to it only selectively. If she was busy mapping, you could wait. She explored to the edge of exhaustion, barely managing to find a cushion or corner before sinking into dreamland. You could pick her up and haul her to bed unconscious like any other infant, all trust and contentment.

She would sleep for hours, limp as a rag doll, then suddenly be seized by a dream, moaning, teeth chattering, feet twitching, eyelids open to show not her eyes but the pink-pearlescent nictitating membrane—a sort of inner eyelid that distributes tears across the surface of a cat's eyes, like a windshield wiper constantly clearing the eye of tiny debris as well as protecting it from infection or scratches. An adult cat's eyes are as big as a human adult's, and she can open the pupils three times as wide. A sort of crystalline mirror behind her retina, the tapetum lucidum, amplifies incoming light by as much as 40 percent. That's what produces the familiar gold-green shine when a cat meets a flashlight beam in the dark.

It is that amplification that allows cats to see in almost complete darkness. They do pay a price for that brilliant advantage, however, some of which you can easily see—the narrow slit that her pupils become in bright light, evidence of her intolerance of excessive dazzle—and some of which you may think a disadvantage but doesn't really matter to the cat, namely, not much in the way of color vision. Another deficit is a meaningful one: Cats really can't focus very well up close. When you extend your palm with a couple of treats as a reward for good behavior or a trick, you may notice that your cat hesitates a moment, sniffing—finding the treat not with vision but by smell.

Augusta spurned store-bought toys in favor of a feather tied to a ribbon. As long as we moved it in just the right way—making it jump from cover, fly from chair to chair, run for dear life—she was thrilled. But if it came toward her, or flew too low or too high, too slow or too fast, there was no game. She taught us to play with her, by her rules. Every toy movement had to emulate wild prey. In this she was not being arbitrary. One of the finest-tuned of cats' senses is their perception of motion. The cat brain's attunement to the slightest gradations of movement has evolved to gauge the flight of prey or the movement of the prey's surroundings that might indicate imminent flight. The cat's visual cortex records images much as an old-fashioned movie camera does—as a fast-moving series of still pictures (faster than the camera, however)—allowing a precise measurement of speed. Hence your cat's ability to intersect the flight of a ball—or of a hummingbird—unerringly. And hence Augusta's utter uninterest in a ball or feather that did not meet her criteria for prey emulation.

A few days after her arrival, the wind known locally as a chinook—the hot dry breath of an exhausted snowstorm—gusted down from the mountains, and in half a morning our snow was gone. Out went Augusta, fearless, tail held high, to map more of her new world. It must have seemed vast, but she was entirely undaunted. We worried that our frisking horses and blundering cows would be big dangers to the tiny kitten, not to mention our case-hardened ranch cats Walter and Penny.

Walter and Penny were big, serious, barn cats. They were not quite feral, but they weren't lap kitties either. Our ranch manager kept a kibble dispenser topped up in the tack room, and on the bitterest winter nights he plugged in a little stove there, so they had a degree of domestic comfort. There was a raw toughness about them as well, however. We had had a number of nocturnal coyote forays that resulted in the deaths of our ducks (three, two, one, none in a week) and then of two ferocious geese. Coyotes had also put an end to a naïve attempt to establish a family of barn cats with a basket of barn-bred kittens. But when coyotes came raiding, Walter and Penny just took to fencepost tops and kept solemn watch. The coyotes knew better than to tangle with them.

When Augusta sniffed her way along the pole fence that served as Walter and Penny's midday observation deck, we were ready to move fast. There was some reason to fear that a pair of long-established cats could well feel compelled to defend their territory, and a territorial attack against a kitten as small as Augusta could be quick and lethal. As she came near, however, and finally passed directly beneath them, Augusta seemed to ignore Walter and Penny altogether, and the big cats, eyes half closed, returned the favor. Undoubtedly both parties were intensely aware of each other, but they did not reveal their awareness in any way perceptible to humans. From then on, Walter and Penny and Augusta were good neighbors, never friendly but always polite.

A continuation of the same pole fence separated the ranch yard from the horse pasture, where we feared she could easily be trampled. The horses were curious, gathering close to snuffle at the furry little explorer, but from how they looked at us and then at her and then at us again it seemed clear that they recognized that respect was due to Augusta. Horses understand rules when they know you are a fair dealer, and it can take little more than a nearly invisible transaction like this one for an agreement to be reached. Horses and cats have a long history of friendship. Cats are not infrequently employed as stall companions for racehorses.

Augusta sometimes sat in the window or on the fence and watched the big beasts at play, but she never entered the horse pasture. The territorial boundary was easy to recognize, since the horses grazed it close all along the fenceline. The treaty required no enforcement. But even had Augusta unthinkingly wandered into the pasture, no horse would ever have harmed her.

The Kitten

ℐ

What had once been a home corral behind our house was now a scruffy, weedy back yard riddled with the burrows of rodents. Augusta's feather-and-ribbon practice probably hadn't been necessary, for from the first she was a highly accurate tracker. She could smell her way along a network of scent marks, dried urine, scat, stray fur, and all the other clues small creatures leave for sufficiently skilled predators to decode. More useful still, Augusta could tune in to the ultrasonic chatter of rodents deep in their tunnels. Cats can hear higher-frequency sounds than any other terrestrial mammal, quite a bit higher even than dogs—up to one hundred thousand hertz (cycles per second). People max out at about forty thousand, if they haven't been to too many rock concerts or ear-splitting bars (which probably eliminates about half of all American grownups).

Augusta's predatory prowess required snowless ground, which we did have intermittently through the winter, thanks to our valley's merciless winds. The unrelenting gusts drove visitors and unseasoned newcomers half-crazy and blew snow into heaping drifts downwind of dead brown-beaten prairie—huntable, if it was not too cold for the princess kitty, for such she was becoming.

But as she matured, Augusta became not only a princess but a deadly killer. She would wait in what seemed rocklike stillness—an illusion, because in fact she was quivering within—till the moment came for the balletic arc that would pin her prey for the fatal bite to the spinal cord. She would chomp a deer mouse down in two crunches—an ounce, give or take. A shrew was one bite, no more than a quarter of an ounce. At two ounces and therefore, I thought, beyond her capability—I thought wrong—a vole was a real chew. Pocket gophers, ground squirrels, pack rats all truly were, at least for now, impossible, but that didn't keep her from crouching flat

7

and switching her tail at the sight of one: the infallible indicator of a mounting urgency to attack.

Augusta herself might well have been prey for quite a few of our neighbors: coyotes foremost, which were always nearby, also foxes, conceivably raccoons, black bears in chokecherry season, and abundant avian predators, including hawks, owls, possibly ravens, and certainly golden eagles. Our hope, perhaps naïve, was that keeping her indoors at dawn, dusk, and nighttime would keep her safe. We knew there were diurnal dangers—rattlesnakes, for example, which wouldn't eat her but surely could kill a curious kitten—but we couldn't imagine denying Augusta her hunting ground.

In the snow, in any case, Augusta was hopeless. It seemed clear that if she had not found her way down our driveway, she would have starved. She simply had no way to cope with snow—her paws were too small to walk on top of it, and she had no idea how to dig. Even in the absence of snow, she just could not stand the cold. When the back yard blew snow-free that winter, I would open the door and she would pad as far as the edge of the porch to survey the possibilities of a return to the hunt; with the first puff of breeze, though, one sniff of ten below, and she was back inside like a shot. We tried a couple of times just plopping her into the snow to see how she would cope. She didn't. She plowed straight to the door mewing in pain, extending her feet to have the ice picked out like a lion with a thorn in every paw.

Once confined to quarters for the winter, which in Montana can stretch into May, Augusta loved to watch water, as long as she didn't have to touch it. A full sink, even better a draining sink, was sheer fascination. She could watch the water in the toilet for hours, and was partial to drinking from it. She loved every kind of falling water. At first she preferred a thin stream from a faucet almost turned off, but over the course of the winter her taste grew more precise. She wanted a drip, not too fast, not too slow. She would slap her paw through it and then lick the water from her toes. Sometimes she was content just to watch the drip, for an hour or more. It was an ancestral image—of an oasis, where a spring often drips in precisely the same way from mineral walls. Augusta was a desert cat.

She was in fact only a few thousand years of descent from the desert wildcat of North Africa. The domestic cat and the North African wildcat were once considered separate species, but they are now recognized as members of the same—*Felis silvestris*. They differ only as subspecies.[1] Unlike the young of any other wild feline, the kittens of *Felis silvestris lybica* can be tamed almost as easily as those of a house cat. They don't look much different either. The big difference is that our cats, *Felis silvestris catus*, now live all over the world, some snuggling on laps, others struggling at the brink of starvation.

The first incontrovertible evidence of something resembling domestication of the modern cat was thought until recently to have been identified in northern China. In 2013 Yaowu Hu of the Chinese Academy of Sciences and a team of fellow researchers published their findings: In an archeological dig in the early agricultural village of Quanhucun, in Shaanxi province, they had found a number of cat remains that—as shown by analysis of the collagen in the bones and by biometric measurement—were, in their view, unmistakably *Felis silvestris catus*.[2]

The researchers also found ceramic grain containers designed to exclude rodents—proof that the people there had a rodent problem, hence a need for cats. The villagers' primary food was millet, and although cats are constitutionally 100 percent carnivorous, the bone analysis also showed that these cats were so hard up that they also sometimes dined on that grain (something today's manufacturers of cheap cat food may be relieved to know). In fact, there

The Inner Life of Cats

was one cat that ate a lot of millet—perhaps a house pet, or a cripple, in any case obviously hand-fed by humans.

The site was carbon-dated to between 5,280 and 5,560 years ago. The ineluctable conclusion is that well before that time there were already domesticated cats somewhere in North Africa or the Middle East. Presumably by trade, those Chinese cats had come more than four thousand miles from the land of their origin.

But a subsequent re-analysis of the Chinese cat bones showed them not to be related to the North African wildcat but another species altogether, a local one, the leopard cat, *Prionailurus bengalensis* (a species that would turn up in the late twentieth century interbred with our house cat species to produce the beautiful domestic hybrid now known as the Bengal, whom we will meet in pages to come). Apparently, the domestication of the leopard cat didn't last, however, because there are no longer any domestic leopard cats in China, only wild ones. All domestic Chinese cats today are the same as ours, *Felis silvestris catus.*[3]

Meanwhile, it was fairly common knowledge among inquirers into feline prehistory that a single cat skeleton had been found some time ago on Cyprus buried next to a human skeleton, dating back about 7,500 years. The cat might have been domestic, or might have been a North African wildcat brought to the island. There had been no further evidence of cat burials there, however, so the single instance was insufficient proof of domestication. Then new bones started to appear in new diggings on Cyprus, in some numbers, including one other cat buried next to a person, carbon-dated at ten thousand years ago. Being buried with a cat did suggest some sort of association more significant than just commensalism (that is, cats hanging around people and people tolerating them for the mutual advantage of rodent control). Archeologists have found that cats were also buried near people in ancient Mesopotamia about five thousand years ago,[4] and cat remains were recently discovered in the predynastic elite cemetery of Hierakonpolis in ancient Egypt, buried there with humans nearly six thousand years ago.[5] Pets or commensals? We just can't tell, at least yet.

The first visual representation of an obviously pet cat—it's wearing a leash—was painted on a tomb in Egypt between 2500 and 2350 B.C. Cats were common figures in Egyptian art for centuries thereafter. The ancient Egyptians loved cats. They worshiped cat-gods. They prepared dead cats for the underworld with the same respect they accorded one another's corpses. One tomb, discovered in 1888, held more than a hundred thousand cat mummies. Sacred cats prowled the temples, and pet cats purred at the foot of Egyptian beds.

(The English are now renowned for their devotion to their pussycats, but in 1888 their ailurophilia was evidently not quite so sentimental. Only one of those hundred thousand cat mummies brought from Egypt to Liverpool was preserved. It is now on exhibit in London's Natural History Museum. The rest were ground up for fertilizer.)

Once cats adopted the Egyptians, they were in civilization to stay. The relationship has not always been smooth, however. Think of all the superstitions about black cats. Cats were often blamed for the Black Plague. Cats have long been associated with witchcraft. Some people even today day are possessed by an angry hatred of cats. Yet cats continue to find their way into human homes and hearts. They live, comfortably, in civilization. But are they civilized?

When Augusta knew immediately to use a litter box and to cover her waste; when she devoted single-minded attention to mapping the minutest details of her new home; when she taught us to play with her by her own rules; when, despite her tender age, she made her own arrangements with the ranch cats and horses; when she hunted and killed and gobbled down her prey; even when she stared at a dripping faucet as though hypnotized—in all these ways our "domestic" cat was acting in patterns set for her by her wild nature. We would come to learn many more, none of which could ever be domesticated out of her.

∽

However wild her ancient nature, Augusta's dependence on us and her affection for us were equally real, and marked her as immitigably domestic. Her species' capacity for love, and the needs that grew from it, emerged only recently in its evolution, and in many individual cats those qualities remain latent and unevoked. Kittens who grow up without human love will, in fact, most likely never show it in adult life. But given the merest touch of a tender hand, the warmth of a lap, the soothing of a voice meant to comfort and calm, even in some rather harsh environments the kitten's innate capacity for human companionship will bloom. They will love, and they will need to be loved.

Like too many other cat owners, Elizabeth and I underestimated the depth of Augusta's emotional needs. The conventional wisdom holds that cats are more attached to their territories than to their people, and that like their wild ancestors, they are essentially loners. Oh, so when she slipped beneath the covers and curled up between Elizabeth's legs, she must have been just seeking physical warmth? When she stretched out on my chest in the morning and purred, it was only because she wanted to be fed?

When we larked off to Mexico for a week to look at birds when she was only six months old, we told ourselves she'd be fine. Worse by far was when we went to New York to be married, and then to Italy, leaving Augusta at home in the care of the ranch manager's niece, whom we didn't know at all. Our cat sitter fled altogether after the first few days, leaving the manager's teenage son, the very type of irresponsibility, to pop in at our house twice a day to leave food for the now nine-month-old kitten. We asked that he also stay a while and play with her a little, but he didn't even pretend to have done so.

When we returned, Augusta wiggled between our legs purring loudly, her tail quivering straight up in joy, and we interpreted her present happiness as proof that she had in fact been fine all along. I know a woman who lived in a bleak little studio and used to go away for long weekends with boyfriends and whose cat, when she returned, always leapt into her arms in delight. She took this as irrefutable evidence that leaving the cat alone had caused the poor creature no suffering. This particular cat did not seem to mind the buildup of waste in her litter box—did not poop or pee on my friend's pillow as another cat might well have done. She ate dry kibble from a dispenser. Video cameras in experiments of parallel conditions usually show the cat looking out the window, perhaps excessively grooming, mostly just sleeping. What are we to conclude? Stick around, we're going to find out.

First let's think back a hundred years, or to anywhere even now where cats are not beloved pets. That would be most of history and much of the world. Where did Augusta's inexhaustible affection for us come from? It was probably an emergent quality, which had lain dormant in her ancestors' inner life through the centuries. Perhaps it had been selected for when cats and people first cooperated on those long-ago farms, and children and cats played, and probably slept, together. And surely it had been cultivated sufficiently in enough households through the generations that it became a reproductive advantage: Affectionate cats were protected, and well fed, and therefore their off-spring were more likely to thrive—more likely, therefore, to produce more successful offspring themselves.

And probably that household affection didn't have to be quite so specific to be sufficient. Probably we've developed a system of sympathy that has created a feedback loop, whereby the more precisely we focus our affection, the more fully, in turn, the cats explain their needs to us ("No, not there, scratch me *there*—ah—yes"). In any case, according to the growing body of scientific understanding of cats' emotional needs, we were doing so many things wrong that we had no right to expect even loyalty from Augusta. And yet she loved us.

And you? When you read about all you haven't done, all you've done wrong, do you also fill with shame? Or are you outraged? *My grandparents on the farm in Vermont* (or Iowa, or Mississippi) *wouldn't even have recognized this nonsense.*

Nonetheless, only a sampling of Kathy Blumenstock's "Ten Ways to Unknowingly Crush Your Cat's Spirit"[6] throws long, black shadows across the universe of guilty cat owners:

Shouting: Raised voices will terrify your cat…

Punishing: Yelling "bad cat," throwing things, motioning in anger, and scolding your cat when she misses the litter box or claws your sofa tells her you are unhappy, even if she has no idea why. Grabbing her and shoving her face in a mess will leave her petrified.…All you're teaching her is to be afraid…

Hurting: Hitting, kicking, physically harming a cat in any way, from a "light tap" to a hard smack, is inhumane, evil, morally wrong, and guaranteed to instill fear in any cat, breaking her spirit, and her heart, in the process…

How about Megan Kaplan's "Top 10 Pet-Owner Mistakes?"[7] At least she brings dogs in, to spread the blame.

Mistake 1: Buying a pet spontaneously.
Mistake 3: Being inconsistent with the rules.…

Mistake 4: Dispensing too many free treats.... Treats lose their training value if your pet gets them for no reason....

Mistake 9: Failing to make your home pet-friendly.... Place litter boxes... in quiet areas throughout your home.... Plug in a night-light beside each one so your cat can find it in the dark.

Megan, had you not heard that they can see in the dark? Still, the night-lights could help *us* from tripping over all those litter boxes.

One more list to tell us how terrible we all, no, most of us, well, some of us, really are: "Ten Things Cats Won't Tell You," by Kelli B. Grant.[8] Again, to be merciful, just a sample.

2. I pretend I'm fine, even when I'm not.... "What does a cat do when it feels good? Sleep. What does a cat do when it feels bad? Sleep."

3. My bad behavior is a result of your bad behavior.

10. I'm not really that funny.

We thought Augusta was funny. After she tore the eight black pipe-cleaner legs off her favorite toy, the Furry Spider, we molded them into the Spider Ball. Our bedroom was long and narrow, with a bare wood floor. Augusta would bound onto the bed, and we would throw the Spider Ball to the far end of the room, and she would pounce on it with such savage glee that sometimes she would turn a full somersault, and then she would grip it in her front paws and try to disembowel it with her pumping back bunny-legs. This was of course an inborn enactment of hunting and killing. And Kelli, I'm sorry, it was *funny*.

And were we inconsistent with the rules? Did we dispense too many free treats? Did I ever yell "bad cat!" when she was ripping the bedspread to shreds and scare her? Guilty. Guilty. Guilty. Did I ever *hit, kick, or physically harm* her? Not really, though I did bop her a few times. Did I do it like a good mother, without wrath? Guilty, guilty.

But as long as we love them, they bless us with their love. It's like people, really. You can make a lot of mistakes, as long as you try to tune in to how your cat understands that you love her. It does require paying attention.

Have you ever seen anybody try to pat your cat on the head like a dog? Rare is the cat who won't flee, duck, or bite. Soon enough, if we pay attention, we learn where our cats like to be stroked, how hard, which direction, with fingertip, flat of hand, tip of nose—and yet we never quite learn where the limit is, whether geographic (not *there*, you moron!) or chronologic (that's *enough* already!).

One reason intolerant people can't tolerate cats is that cats express themselves like cats. Dogs really do try to figure out what you want to see and hear, and they deliver it—good dog! Cats love for you to love them, but they're just not wired for tail wagging, slobbering, and joyful woofing.

So what *are* cats trying to tell us? There is recent evidence that they and we share emotional centers in the brain of striking similarity, and that cats' vocal expressions originate in those areas. Could it be that we recognize what they're saying in ways we don't consciously understand? Pascal Belin, of the University of Glasgow, and a team of other researchers were interested in Charles Darwin's hypothesis that a common set of mechanisms underlies the expression of emotions in both humans and animals.[9] The scientists used functional magnetic resonance imaging (fMRI) to examine activity in the brains of several different species—including people and cats—in response to a range of emotion-related vocalizations. They employed positive ones, such as expressions of satisfaction, sexual pleasure, and hunger, and negative ones like alarm calls, distress signals, and cries of pain. Their first objective was to see if the same regions of the human brain were stimulated by these vocalizations as the brain regions of the creatures in the act of expressing themselves—and they were.[10]

Because it was possible that the humans' familiarity with cats might predispose them to an understanding of cat language, the researchers also used vocalizations of rhesus monkeys, which neither the human subjects nor the cats had ever heard. The emotional expressions of both the cats and the monkeys, positive and negative, showed precisely the looked-for activation in the humans' brains, and the human brains responded accordingly to the negative and positive emotions. (This result at least hypothetically demonstrates a line of emotional language development in mammalian evolution, dating way back before our hominid ancestors started actually talking in words.)

But the humans turned out not to *realize* how the cats' and the monkeys' emotional "speech" rhymed with their own. They might have understood some of the cats' expressions, from their own experience with cats—though still not all that much—but they didn't get the monkey talk at all. What their brains knew, they themselves didn't know.

Perhaps someday we will be able to dig deep enough beneath our consciousness to sense what our brains are already processing. Maybe even now some yogi or other mystic could do it. In the meantime, if we listen well enough to what our cats have to say, we can understand quite a lot of it. More comes to us year by year from science.

Beyond the nice lists and internet advice whose authority one never really can know, there's now solid, tested, scientific evidence for what we can do to improve our cats' quality of life (or assure us that what we're already doing is good). But a great deal of it has not yet made it to the public at large, or is intermingled with erroneous information that gets passed along from cat book to cat book without their authors bothering to check whether it's true or not. The first great collection of scientific studies and essays, *The Domestic Cat: The Biology of Its Behaviour,*[11] was published only in 1988, and many of its findings have yet to reach the majority of cat owners.

The frustration the scientists felt about their work emerged in one of the essays, "Practical Aspects of Research on Cats," by Claudia Mertens and Rosemary Schär:

> Veterinarians...often lack scientifically funded knowledge of a) normal cat behavior, b) the minimum living conditions "acceptable" to the cat, c) the causes of behavioral disturbances, d) what humans expect from their cats and the ability of cats to satisfy those expectations....The scientific basis of our ethological, ecological, sociological, and even psychological knowledge of domestic cats and the human–cat relationship is limited, and many of the existing studies summarized in this volume have not been accessible to veterinarians or the interested layman. On the other hand, popular cat books generally are not written by scientists and most often are not up-to-date (to say the least).

The last statement is still true, alas, and it's still true that far too many veterinarians are not well trained in the full spectrum of cats' needs. But just in the last few years that has been changing fast. A landmark document from the veterinary community[12] opens with these words: "A cat's level of comfort with its environment is intrinsically linked to its physical health, emotional well-being, and behavior." As self-evident and commonsensical as that may sound to many of us, it is a very different order of values from those held by veterinarians only a generation or two ago—indeed from many cat owners' ideas even now of what cats need. There are all too many cat owners, at this moment, whose idea of care is feed 'em, house 'em, take 'em to the vet if they break their leg, and otherwise, *ptui*—they can take care of themselves, for God's sake, they're *cats*. That's why I didn't get a goddam dog. You don't have to walk 'em, they don't slobber all over ya, they poop in a sandbox, you can leave 'em alone for any old time as long as they get fed.

This set of attitudes isn't a myth. A well-conducted study confirms that dogs get much better veterinary care than cats, and significant numbers of cat owners don't even believe cats can get sick.[13]

We put human beings in solitary confinement, after all. Slide their dinners through a slot, award sunshine, maybe, once in a while, for docile defeat. We put our own parents in "long-term care facilities," to be strapped down, downed out with antipsychotics, scrubbed down like doomed livestock on the slaughter ramp. And funny thing? Recently some prisons and old folks' homes have brought in cats, and even psychopathic murderers and previously unresponsive wanderers-in-fog begin to smile, touch the cats, cuddle them. Sometimes something breaks loose inside them and they weep.

༄

An American psychologist based in Switzerland, Dennis C. Turner, has studied the effects of the companionship of cats on the emotional states of people living alone with cats, people living alone without cats, cat-owning couples, and couples without cats. Animals, science generally claims, don't have the same rich inner lives that we humans do. This, of course, is anathema to many pet owners. Turner and his colleague Zana Bahlig-Pieren decided to test the widespread belief of dog and cat owners who believed that they understood their pets' behavior, emotions, and intentions.[14] The experimental subjects included roughly equal numbers of people who had experience with dogs or with cats or with both, and those without any. They were shown a sequence of photographs and videos of the animals and then asked to answer a series of multiple-choice questions. For each question there were three anthropomorphic answers (answers that "ascribe human mental experience to non-animals") and three scientific answers. Each of the three answers was rated by three ethologists, independently, as "very plausible," "plausible," or "implausible."

The people who had experience with both dogs and cats scored highest in scientific ("ethologically very plausible") understanding of the scenes. Dog people were more anthropomorphic than cat people, but both still showed a high level of understanding. Even the people with no personal experience with dogs or cats at all were pretty good at interpreting the animals.

John S. Kennedy, in *The New Anthropomorphism*, published in 1992, vigorously condemned any use of anthropomorphism by scientists, but according to Bahlig-Pieren and Turner he also maintained that, "the capacity for anthropomorphic thinking may be a product of natural selection, probably because it proved to be useful to our ancestors for predicting and controlling the behaviour of animals." Which kind of sounds like trying to have it both ways. Whatever—it does seem pretty clear by now, on the basis of a good bit of sound science, that at least some of our sense of understanding of Augusta was not just wishful projection.

Still, there was so much we did not understand. That first winter was very cold—it hit thirty-six below one night—and so Augusta's life was an indoor one. Why did she flee at the sound of most men's feet from the moment they came in the door but would trot right out to greet our one friend who was six feet eight inches tall and wore size fifteen boots? How could she settle gently in the crook of an arm, or lie curled in the curve of a neck on the pillow, but resolutely refuse to sit on a lap?

Another friend, a longtime naturalist practiced in close observation of all sorts of animal behavior, asked, "Why don't you teach her to fetch?"

Cats can fetch? Many a morning we would throw the fuzzy Spider Ball down our long skinny bedroom and Augusta would give chase, as any cat would do, then wait for one of us to go get it, go back to bed, call for Augusta to follow, wait for Augusta to figure out that she was being summoned, wait for Augusta to jump back up on the bed, and throw the Spider Ball again. But finally, one miraculous day, she brought it back, sort of, at least to the foot of the bed, to cheers, huzzahs, *good kitties!* and a single crunchy treat. From then on we brought her gradually up to about 60 percent success as a retriever cat, at which point her boredom, or ours, or both, intersected her rising retrieval curve and flattened it flat. Probably that was just a pause and, if we had kept up the good work, she would have gotten better, but all three of us were, frankly, sick of the Spider Ball. Augusta's future as a trick cat went dark.

Spring in Montana begins as melt, mud, overcast, and thunderstorms, and then one day the world is green, wildflowers spangle the hills, and your cat is bounding across the ground squirrels' lawn, a black wiggle monkeying up a spruce and gone. Gone all day. Had an eagle gotten her so soon? It was too cool for rattlesnakes yet; too bright for coyotes; the chance of a cow stomping her were slim; the horses were her friends; my ranch partners' dogs were harmless Labs. A bear? Old-timers assured us they didn't come down till the chokecherries ripened, hot summer. We were a long way from the road, and there was hardly any traffic anyhow. It was a good, safe place to be a cat.

The ecosystem was swirling and spinning into order minute by minute, first algae clinging to cold rock at river's edge, first grasses slicing up through winter straw, first frog eggs bursting into tadpoles in the pond, eggs everywhere, tiny green ones curling new leaves around their sticky fuzz, pearlescent butterfly eggs disgorging caterpillars, magpie eggs cracking from within at the insistent tap of naked black blind chicks, chartreuse willow shoots swelling with yellow catkin buds. The deer mice in their hundreds were setting up housekeeping, the dominant males marking territory with urine and all their followers racing about, checking, checking to be sure they were getting the system straight. Augusta followed the re-establishment of mouse society with close attention, in no need of schooling. Augusta's gift was evolution's gift.

Her sense of smell was uniquely attuned to this kind of signal. When the complete genome of her species was published in 2014, one of the more remarkable findings was that the gene "repertoires" of dogs and cats have evolved to differ in a particularly significant way: As is well known, dogs are terrific smellers, exquisitely sensitive to what researchers call odorants; but in the same gene region of cats, the repertoire is specific to an equally exquisite sensitivity to *pheromones.* "These results," the researchers write, "add further evidence supporting cats' extensive reliance on pheromones for sociochemical communication."[15]

When your cat looks at you in a puzzled, dreamy way with her mouth hanging open and her upper lip pulled up, that means that she has been using her "booster" sense of smell, which requires her to open her mouth in this oddball gape expression, usually known by its German name, flehmen. Behind her upper front teeth are two tiny tubes, each of which leads to what is called the vomeronasal organ, a sac containing an array of at least thirty exquisitely sensitive chemical receptors.[16] Cats displaying flehmen are usually engaged in a subtle reading of scent marks left by other cats, and they are in a state of almost hypnotic absorption.

All that pheromone gab doesn't go on just between cats. Augusta could read the deer mouse community map like Leonard Bernstein scanning a Beethoven score. Her perception of spring was not like ours. We see the world around us and we glory in its multiplicity of wonders. The cat tunes in to an invisible web of scent, of creatures she is designed to kill.

Montana summer is all *allegro.* The sun may paint the northern sky red till midnight: Augusta's first summer solstice never got truly dark, for the moon was a waxing crescent, and even when the moon is new, unless the clouds are thick indeed, the stars over Montana are sufficient to illuminate a path. At 2:52 A.M. the sun began to brighten the sky, albeit dimly. Fog filled the valley, and a thin, steady rain was pittering on the roof. Augusta sat on the windowsill disconsolate. She hated rain.

Outside, it looked as though every animal visible also hated rain. Magpies hunched in the spruces, soaked and cold. Two golden eagles brooded on fence posts, disheveled and not on the lookout. The famous big sky hung heavy and silent. The horses stood nose to tail, heads down, glistening eyelashes closed. The cows found a lee slope and turned their butts to the wind and did what they always, reliably, in all weather, did: eat. Leafblade and root were in joy, the soil still warm from yesterday's long sun, their growth rate astonishing, as was that of the cows and their scampering, mud-flinging calves. Buds were splitting, petals slick with light, mayflies new-hatched clinging to the underside of streamside leaves. When the midday wind blew the clouds apart, the whole orchestra swung into its principal theme, at a storm-driven pace.

In fact already the late-day-to-come's black thunder was grumbling in the mountains, with the impatience of an omnipotent monster. Augusta sprang through the tall wet grass like a stotting mule deer and disappeared among the granite boulders behind the house, some of them tall as the house, on her own, free.

She returned when called, drenched and shivering, and loved the rough toweling we gave her. Her underfur didn't amount to much—truly she was a desert cat—and when she was dry she was soft and fluffy as a baby toy. When I ran my hand down her spine to her tail and her tail in reply lifted to flag-high, we felt each other's full attention. I didn't know at the time that the sweet spot just above the tail is a scent gland (humans can't smell it) and that my fingernailed trace across it, picking up a mark from her, was a sort of kiss. She closed her eyes, slowly opened them, and between her black lips gave a quick downward flick of her pale pink tongue— gestures of affection. She smelled so good I wanted to bury my face in her, which she recognized as a genial howdy and reciprocated by sniffing around the hem of my pants and the soles of my shoes. Satisfied then, she took up her finely detailed grooming, one back foot held high. A few tongue-strokes up the back of the thigh, then a chew on the belly, then a licked-paw half-face wash—was there the slightest order to it? We pattern seekers are tempted to see one, but there wasn't, and there isn't.

The afternoon closed with a grand finale, the sun engulfed in blackness and then the earth-cracking, gut-pumping *fortissimo* of the full storm, sheets of rain, hammering hail, wind in waves across the pastures, then sudden stillness and the thunder suddenly far, echoing out across the prairie north. Meadowlarks resumed their liquid arias, and yes, there was a rainbow, double in fact, in the east. The cat had slept through it all.

As the bright days hurried by, Augusta grew rapidly more independent. Her forays out among the rocks and sagebrush lasted longer. We reassured ourselves that she was too big and claw-capable now for a red-tailed hawk (our commonest daytime raptor) to try to take. An eagle was possible but somehow seemed unlikely so near the houses and barns. The dangerous predators were all, at least so we believed, twilight dwellers—the fox, the bobcat, the great horned owl. If one of them happened to break that rule, as a coyote might, surely Augusta would shoot to the top of a willow or aspen or big rock out of reach, as we had seen her so easily do. (She had still not mastered coming down from high in a tree, but word was that any cat would figure it out if left up there long enough. We did hope so: Our local fire department, all volunteer and twenty miles away, would not have mobilized eagerly for a cat-up-a-tree call.)

And proud she grew, too, the little huntress. Like all her kind she brought gifts home. Sometimes they were fully alive and so much fun to chase under the refrigerator (where her staff could be counted on to shoo them out somehow, with wooden spoons, broom handles, straightened coat hangers) or to bat from chair top to lampshade, bleeding only a little; and sometimes plumb dead. She would drop a deer mouse at our feet, give a *mrop!* of victory, bite it in half amidships, and disappear it in two bites, chomp chomp and swallow.

Now sometimes she did not come to the call of her name. Now sometimes she would stay out even into the long shadows and chill that tell the always shockingly early onset of autumn in August in Montana. I'm hunting, don't bother me. But we knew that as dusk came on, danger was real. We called and called, and finally there she would come, bounding through the grasses now two and three feet tall, black shape and bright eyes for an instant, then a dip in the green, then her eager face again, indefatigable, full of joy, straight into the bright kitchen, *Where's my dinner?*

Then came the time she did not come. It was an acutely inauspicious time for her not to come: At the base of many of the giant boulders, where runoff rain collected, chokecherry bushes grew and the chokecherries were ripe and black bears had come down from the mountains to feed on them. There's not much cherry on a chokecherry, a leathery skin and little more than a thin squish of fruit, and the fruit pretty sour, but the bears went crazy for them. They lost all their customary sense of where they ought not to go. As soon as the sun was good and down, the bears would pad through the riverside willows and up along the corrals and between the barns, just under the lighted windows of people (us) having dinner, listening to music, watching videos (we didn't have satellite TV yet, much less cable), laughing, and in a nervous or accidental pause you'd hear them, their claws scraping on the tin roof of the root cellar, the snap of whole branches breaking as they ripped them down to reach the cherries at the tip, sometimes the *mawww* or bark of a mom scolding her cubs—the *truly* dangerous bear, that mom. You damn well stayed inside. But Augusta was out there amidst them.

It wasn't night yet, yet. We called and called. We ventured as far as the fence with flashlights and called. I, having written a book about grizzly bears, supposedly knew that black bears could be scared away if you took a strong enough tone with them. But: I had met a black bear mother and cubs once on a trail in Idaho and they had not gotten my message. The mother had sent the cubs up a tree and then turned to me, snapping her jaws, a bad sign, and stamping her front feet, another bad sign. By the grace of chance there was a footbridge not far behind me, and I took one very, very slow step backward toward it, saying very, very quietly something like, "Now, you know, Mrs. Bear, I have only the most peaceful of feelings toward you." And as I reached the bridge, which smelled of creosote and blessed humanity, she snorted at me, then woofed at the kids up in the tree, who came scurrying down. Family assembled, they ran like hell away.

Plus, chokecherry bears are not in their right minds. So. I was scared. But if I could help it, I wasn't going to let Augusta be eaten by a bear.

Hence, stupidity having taken over, I ventured forth, flashlight in hand. (I did not have a firearm—that's another discussion.) Calling, quietly now, Augusta, Augusssta. Between those big rocks by now it was really quite dark. Then I saw, in my flashlight beam— this all happened very fast—the eyeshine at just the right height and just the right narrow distance apart to be the eyes of a bear, walking placidly not at me nor away, just along, and then, behind, that, the eyeshine of, oh, God, no, a cub, and then another cub, and then, smaller, a little behind but not far, a cat.

Augusta was out walking with a black bear and her cubs.

I did not know what to do, so I did nothing. For a long time I just stood there in the dark. Then I went home. Augusta arrived shortly thereafter.

Chapter Two

Becoming a Cat

A lot of cats have mental problems. Augusta certainly did. Sudden loud noises sent her diving for cover. The only strangers she would even approach were those few quiet women long familiar with cats who had the patience to sit on the floor, speak softly, and wait. Even then, it might take a while for Augusta to slink from hiding and sniff the lady's extended forefinger. Even then, one twitch and she might be gone like a bullet. Men, almost universally, she had no tolerance of whatever, her exceptions being me and our aforementioned six-eight friend with feet like anvils. All her life long she never sat on a lap. She would sit close by, purring and content, but laps, no, those were danger zones. She would sleep on our feet, and, later, between Elizabeth's legs—higher up the colder it got—or on top of your head, but as soon as you sat up you were scary.

We were fortunate. Such neuroses as Augusta's were, comparatively, piffle. Thousands of cats won't pee in their litter boxes, preferring to squirt the walls or furniture (and the stench can be ineradicable). Furniture, oh yes, thousands of cats rip it to ribbons. Thousands scratch or bite. There's yowling all night. Wool sucking. Eating houseplants, including poisonous ones. Refusing to eat. Obesity. Running away. Fighting. Compulsive vomiting. Compulsive self-grooming. Feline hyperesthesia (a sort of epilepsy, in which the cat attacks itself). Aggression toward all humans. Terror of all humans.

It's tempting to ascribe many of these difficulties to the domestic cat's essential wild nature in conflict with its life of confinement, or to our own misunderstanding of our own mysterious pet, or to neglect or maltreatment that we didn't even know we were guilty of at the time. In many cases, however, the trouble has started much earlier, before you ever laid eyes on your cat.

When things go right for a kitten, it is almost certain to grow up to be a sociable, even-tempered, happy cat, even in difficult circumstances. But what does "things going right" mean? Amazingly, in all the years of cat domestication, there had been no fully scientific study of that question until late in the twentieth century. There was a great deal of folk wisdom, some of it accurate, some of it not. There have been dozens of books on how to raise cats, some of them sound, some of them not, much of their guidance not sufficiently clearly defined or sufficiently widely tested to yield reliable outcomes. Seemingly as simple a question as the right age for the adoption of kittens seems not to have been addressed with much consistency in veterinary schools. (It would at least have been agreed that abandoning a tiny kitten in the snow in Montana fell short of the ideal.)

The one study that demonstrates nearly everything we need to know about the ideal early life of the domestic cat has lain essentially buried for almost thirty years. The study is not entirely unknown—some popular writers have referred to it—and some of it simply confirms old common sense; but its essential discoveries were revolutionary.

The text of the technical article "The Human–Cat Relationship" by Eileen B. Karsh and Dennis C. Turner never made it to the internet. The title occurs only in a few other papers' reference lists. The study's sole public existence is as part of a collection of papers from a symposium called "Cats '86—The Behavior and Ecology of the Domestic Cat," which was held at the University of Zürich-Irchel in Switzerland in September 1986. The collection was published in book form by the Cambridge University Press in 1988 as *The Domestic Cat: The Biology of Its Behaviour.* It was not a best-seller. As far as I have been able to discern, there has been no significant specific follow-up to the Karsh and Turner paper.

Which is a shame, because "The Human–Cat Relationship" is almost certainly the single most important study in all the scientific literature on the domestic cat. The information in it has the capability of making a great difference in the number of kittens that grow up to be happy, well-adjusted cats, and, thereby, a great difference in the lives of many people.

It took me months to find Eileen Karsh. Nobody in the Temple University psychology department remembered her. I could find no email address, no phone number, no scholarly references other than that single study. She didn't exist on the internet. I finally found a list of donors to some charitable cause in Philadelphia that included "Eileen K. and Lynn Hammond," which mentioned that Dr. Lynn Hammond had also been a member of the psychology department at Temple. There was another emeritus Temple professor of psychology on the list, and eventually I was able to find his email address. He told me that despite the name more often given to women, Lynn Hammond was a man, and that his wife, Eileen K. Hammond, was probably the same as Eileen Karsh. But he hadn't seen his former colleague in thirty years, and he had no idea where he was. I started searching for Lynn Hammond in telephone listings in and around Philadelphia and finally, bingo, there was one in the suburb of Wayne, Pennsylvania. I called, and a woman's voice answered, with an old-fashioned New York accent. Was this by any chance Eileen Karsh? It was.

Would she be willing to talk to me? She started talking, and I could hardly get a word in edgewise. I said I wanted to come and see her and have her tell me about her research and her whole career. She said she'd be delighted. So I took the train down from New York to Philadelphia and a creaky commuter train out the Main Line to Wayne. A big blue BMW jounced into the station parking lot.

She made me promise not to write about her house. I couldn't imagine why, and I said I was interested only in her work anyway. What I can say about the house is that there are a lot of cats in it. Karsh was surprised but gratified that I had found her work. She knows it has never gotten the attention it deserves.

Eileen Barbara Karsh was born on July 6, 1932, in New York City, and grew up in the verdant Riverdale section of the Bronx. She excelled as a student at the private Fieldston School, and she went on to Smith College, where, she later discovered, she fell within the 5-percent quota of Jewish girls. She dropped out after two years to marry, got divorced after three years, and then entered Barnard College, where she developed an interest in psychology. Her father, a physician, wanted her to go to medical school, where there were very few girls of any stripe, but Eileen wanted to pursue her psychology studies in graduate school, where there were even fewer. Columbia, Harvard, and Yale all accepted her. She chose Yale. Both the students and faculty were all male. She had counted on a small income from a teaching assistantship, but it was Yale policy that a woman could not hold one. The psychology department was not congenial to its only female graduate student, but she persisted in her study of animal learning—using lab rats, as nearly everyone in the profession did—and in 1959, with a dissertation on the nature of reward and punishment, she achieved her Ph.D. The University of Pennsylvania, where she went for postdoctoral work, was no friendlier, but she managed two years of research there nonetheless, with a National Institutes of Mental Health grant to look at areas of the rat brain that controlled feeding behavior. She discovered that the stress caused by forcing her rats to swim caused damage to their hypothalamus.

For the first time since childhood she got a cat, a red tabby named Cyrus, but her work continued with rats, first at Swarthmore College, from 1961 to 1963, and then from 1963 to 1967 at Drexel University, where she was appointed to an assistant professorship in biomedical engineering, specializing in statistics. Finally she found her permanent academic home at Temple University, in Philadelphia, where for the first time she was fully employed as a professor. "*Eight years* after I got my Ph.D. it took to get hired by a university psychology department." Things were tough in academe for even the most brilliant of women.

In 1968 she married fellow psychology professor Lynn Hammond, who suggested that she might prefer working with cats. "He was right," she says. "I didn't feel like shocking rats anymore.

"I got a small grant, trying to do a sort of Pavlov thing with the cats, a progressive denial of food, trying to create an experimental neurosis. It didn't pan out, because the kittens were too old. I began thinking about how a lot of mother cats don't help their newborns if they can't reach a nipple—they just die. And so I developed an interest in early-stage relationships. Attachment. I took very good care of my cats. Except for the very young litters, who had to be kept in a cage, I always let them just move around the lab, which they enjoyed. They enjoyed each other's company, and mine.

"Cruelty to lab animals was the norm back then. Cats and other lab animals, when the researchers were finished with them, would be sent to med schools for—for a fate worse than death. Really, worse. There were cruel experiments, yes, but also there was actual sadism.

"Lab animals now are pretty happy—happier than animals in zoos, certainly. They develop relationships with their people, they get used to their routines. A lot of people don't believe this. When I got going with a larger number of cats—and they were always happy, very happy—people were protesting outside. I was on television, and I was not nicely portrayed. I was the villain.

"And then people in the department, some of them, thought I was 'playing with cats,' because they weren't in cages. Sometimes, if somebody opened the door, a cat would jump out, and they thought the cats were trying to escape some terrible situation. If you know anything about cats, you know that if you open a door with a cat on the other side, he might just jump through it. I mean, how are you going to study *attachment* between people and cats if the cats are in cages?"

Karsh's interest in early-stage attachment was the key that unlocked mystery after mystery. She was fascinated by the work of a biologist named Zing-Yang Kuo, who had published research in 1930 showing that "when kittens were reared with rats, they grew up to be cats that did not kill their cagemates and usually did not kill other similar rats."[1] In later work, Kuo raised kittens with puppies, rabbits, birds, rats, and other kittens, in various combinations. He measured the degree of the kittens' attachment to the other animals by removing the latter from the study enclosure, leaving only the one kitten, which "would become restless, cry, and look as if it was in great distress. If the kitten lived with five puppies, as soon as he put in a puppy, the kitten calmed down. However, when the kitten lived with four other kittens and one puppy, the kitten remained restless and distressed in the presence of the puppy until another kitten was put in the cage."

Michael W. Fox, in a 1969 experiment, had raised some kittens with puppies starting at the age of four weeks, and with other kittens starting at the age of twelve weeks. The kittens that started early played easily with the puppies and never showed any fear of dogs. The kittens that had never seen a puppy until the age of twelve weeks were afraid of them, and remained irrevocably so.

There had also been earlier studies of the effects of human handling on infant animals—baby rats as well as kittens. Both species initially reacted with distress, but both also showed faster rates of physical development. In one experiment, Siamese kittens were handled for twenty minutes every day for the first thirty days of their life, and "they opened their eyes one day earlier, emerged from the nesting box 2.6 days earlier, and developed Siamese coloration earlier." Later analysis suggested that the more rapid development may have occurred because the babies' cries for help stimulated their mothers to pay more attention to them—which had been shown in other research to have similar effects.

Another study had demonstrated that even very brief handling—as little as five minutes a day—caused kittens to approach people more readily and more often. Another experiment added play to the handling, as well as varying the number of people involved. The kittens that played with five people grew up to be the least afraid of strangers, while one-person kittens became more affectionate.

This basket of scrambled and inconsistent data was all Eileen Karsh had to go on when she started her work on cats. Her first experiments focused on confirming earlier research, particularly the studies that investigated the effects of handling young kittens. She found that "eight cats that were handled for five minutes daily approached a person faster (32 seconds) than seven non-handled cats (128 seconds)."

She also considered earlier, rather horrific studies in which infant animals were separated from their mothers and from all other contact. In one, kittens had been kept in complete isolation up to the age of ten months and then exposed to dogs, rabbits, rats, guinea pigs, canaries, parrots, sparrows, and other cats. In all cases—big surprise—the poor kittens' response was "predominantly hostile and attacking." Another god-awful experiment removed three groups of kittens from their mothers at the ages of two weeks, six weeks, and twelve weeks—and kept them away—and then looked at the results when they reached nine months. Another non-surprise: The cats that had been kidnapped at two weeks "displayed excessive undirected activity, disorganized behavior, fear of novel situations, and inability to tolerate delay in feeding…[and] showed aggression and non-cooperation in a food-competition contest, while cats from the other two groups cooperated."

Similar torture had been inflicted on puppies, with parallel results. Horrible as all this had been, as the data piled up Karsh began to see patterns defining narrow windows of time in their infancy when kittens were particularly sensitive to socialization— first with their mothers, then with their littermates, and then with people. There had been a few vague speculations on these time periods, but never any scientific attempt to define their boundaries or their nature.

Suspecting that socialization with people may have been possible at an earlier age than had ever been thought, and that it may have extended much later than had been widely believed, Karsh devised an experiment that divided her British shorthair kittens—all born in the Temple labs—into three groups. The first would begin to be handled at the age of three weeks, for fifteen minutes every day; the second, at seven weeks, "about the age represented in most pet care books as the best time to begin handling"; the third group would not be handled until they were fourteen weeks old.

"Handling" is a somewhat crude word for what Karsh and her helpers did, which was to hold a kitten in their laps and gently pet and stroke it, and then return it to its mother and its brothers and sisters. Each kitten was handled by four or more different experimenters. All of the kittens lived with their mothers and littermates in large cages up to the age of eight weeks, after which they were all free to roam around the several rooms of the lab.

From the age of fourteen weeks up to one year, they were tested for "friendliness or attachment" every two to four weeks. Karsh had two ways of measuring that. In the first, one person put the cat in her lap while another timed how long the cat stayed there until it jumped down. In the second test, "A person was seated at the far corner of the room on the floor, with a chalk line drawn at a distance of fifteen centimeters around the person. The cat was introduced and the time taken to reach the person (latency) was recorded. Trained observers also recorded time spent with the person within the chalk line during a three-minute test period and other friendly responses such as head and flank rubs, purrs, and chirps."

The earliest-handled cats stayed in their handlers' laps for an average of forty-one seconds. Those that weren't handled until seven weeks stayed for twenty-four. The ones who were untouched until fourteen weeks stayed only fifteen seconds.

The approach test had startlingly different results. The early-handled kitties approached within eleven seconds, but the middle group and the lates were almost the same—thirty-nine and forty-two seconds. Clearly, seven weeks was too late to start overcoming the kittens' shyness. This was a major finding.

Augusta's personality comes vividly to mind. Wouldn't sit in our laps, and a stone fraidy-cat.

Karsh found the responses of the three-week-old kittens so striking that in her next experiment she included a group that were handled starting at the age of one week. These are really *little* kittens. Usually they've just opened their eyes the day before—and their eyes are always blue at this point—but they still can't focus. Their ears are just starting to fold open. They weigh barely eight ounces. They're unaware of their surroundings: Their whole world is mom and their littermates. And yet handling them in their earliest infancy made a significant difference in their later lives. When the tests began at the age of fourteen weeks, these kittens stayed in laps longer than those started even at three weeks. They approached people more quickly as well.

Now Karsh wanted to improve the precision of her findings. In the previous studies, the kittens had been handled for differing lengths of time, and perhaps that had distorted the results. In the new study, she would have a larger sample size—seventy-five kittens in all—and each group would be handled for precisely four weeks: from ages one week to five weeks; two to six; three to seven; and four to eight.

She was looking for precision, but suddenly she thought of a possibly serious confounding variable, which could in fact have been screwing up her experiments all along. She was thinking about Pavlov again, and how back in 1927 he had observed that the dogs in his experiments fell into two broad natural character types, which he called excitable and inhibited. The excitable types weren't easily conditioned. Other researchers had done similar classifications of dog personalities by breed, and there were informal notions about cat types: "Siamese cats," wrote Karsh, "were described as outgoing, demanding of attention, and vocal, while Persians were described as lethargic, reserved, inactive, and not desiring close contact." Now, on the hypothesis that some cats, like some humans, are born shy or born outgoing, she added assessments of timidity versus confidence to her approach tests, while her co-author, Dennis Turner, in Switzerland, using different methods, measured what he called shyness versus friendliness. Whether or not these traits were innate, they seemed to be stable from early kittenhood throughout a cat's lifetime. One strong possibility was that since cat litters commonly have multiple fathers, timidity or confidence might be inherited from the father. Karsh recalculated her data to include timidity and non-timidity of all the kittens, as assessed by three of her assistants. Sure enough, once the timid cats were excluded, the results were even stronger:

"The sensitive period of socialization for cats to people is from two to seven weeks of age."

This was the gold. It meant that kittens should be petted, cuddled, talked to, and taught the sight and smell of humans much earlier than anyone had ever known, and that the period when handling is effective tapers off much sooner. Breeders, shelter staff, veterinarians, self-appointed experts of every kind had had it wrong—and for the most part they still do.

It took a great many kittens to sustain Eileen Karsh's research, and as they matured they were no longer of use in the laboratory. Those who had been hand-raised from very early ages were supremely adoptable—friendly, confident, unafraid. Even the ones classified as shy were still way ahead of their peers in shelters and, God forbid, pet shops. And of the fourteen-weekers—the kittens deprived of petting and play till the age of fourteen weeks—Karsh now says, "Of course they'd be a little harder to tame, but it's not like they were damaged in some way. A little gentle care would bring them around. I mean, most of them, by the time we adopted them out, had been with us for a while, running around loose with me and my assistants in the lab. It was a pretty good life. At first we gave them away for free—a good British shorthair like that was worth three hundred dollars—but we felt that people were sort of taking them for granted, so we started charging thirty-five dollars, which was just the cost of neutering. Naturally I kept a few for myself."

A piece headlined HER JOB'S THE CAT'S MEOW in the Philadelphia *Daily News* described the Temple cat lab thus: "Cats roam freely, snooze on roofs of open cages, or enjoy various climbing poles, perching shelves, and cat amusements....The 'lab' seems more like a combined feline fun arcade and cat hotel." Karsh was quoted: "Most of the time when you see cats in the laboratory, they have electrodes in their heads."[2]

The serial adoptions offered a perfect opportunity for more research on attachment. Dennis Turner, in Switzerland, had already looked at the question of whether the mother's presence was a significant influence during the early period of contact between people and kittens. He found that when the kittens were very young, having mom nearby made them somewhat bolder in entering the open test room—but they would then run not to the testing person but back to mom. A few weeks later in their lives, mom's presence seemed to give them confidence, and they would approach the person more readily than kittens whose mothers were not in the room.

Both Turner and Karsh found that kittens reared in people's homes tended to develop friendlier and more trusting personalities than even those with the most handling in the labs. It proved impossible to measure the time the home-reared kittens spent with people, and the number of people interacting with them was equally difficult to pin down, but it seemed a safe bet that a home would be a richer environment than a lab. More people spending more time with their kittens—that was the best of all possible ways to rear a happy, bold, and sweet-tempered cat.

Turner tested the question of whether feeding, alone, could be a significant factor in establishing or maintaining a sense of attachment in kittens or cats with the people who fed them. The experiments were somewhat elaborate, but the results were simple: "The act of feeding a cat can enhance the establishment of a relationship, but it is not sufficient to maintain it. Other interactions (petting, playing, vocalizing, etc.), are required to cement a newly founded relationship." To a sensitive and sensible modern cat owner this finding may well be met with a "Duh!" but it must be remembered that there are many other owners who think feeding is about all they need to do to keep their cats loyal—and who then, dismayed, characterize their cats as "aloof" and "standoffish."

Eileen Karsh was interested in how people chose which kitten they wanted—and in how the matches might be made better. She was immediately struck by the fact that physical appearance seemed to be almost the sole criterion. Sometimes people were looking for a cat to replace a previous one, and wanted a look-alike. "One woman told [me] that only the personality of the cat was important to her. When she was shown the first few cats, she remarked that she couldn't relate to a cat that was all white or all black. When questioned about this, she responded that she related better to cats that were gray or striped."

It didn't take long for Karsh to conclude that just describing how people made their choices wouldn't make for a very satisfying study. She wanted to influence how they chose—with the goal of both happier adopter and happier adoptee. She had already conducted a long-term study that showed that a successful match between cat and owner, especially an older owner, could have remarkable benefits for the person.

Eileen Karsh's study of the effects of cat ownership started with interviews of twenty people, with an average age of about sixty, asking them to rate themselves on scales of loneliness, anxiety, and depression.[3] They all came out pretty much the same. Then seventeen of the twenty interviewees adopted cats. Of those, eleven kept their cats for more than a year. Over the course of several follow-up interviews—of the long-term cat owners, the short-term owners, and those who had not adopted a cat at all—Karsh found that at the end of a year, the long-term cat owners were significantly less lonely, anxious, and depressed. After two years, four of the long-term owners who had had seriously high blood pressure at the beginning of the study now had much lower pressure; one was able to discontinue medication entirely. Two who were diabetic showed lower blood sugar levels. No such changes were seen in the subjects who did not own cats.

More recent, generally broader studies of the effects of pet ownership on human health have shown varying results, and there has been wide disagreement among scientists. Many have found significant benefits—physical, emotional, social, and psychological.[4] Dennis Turner has written, "The cat might be an ideal co-therapist for clinically depressed persons. Psychiatrist Daniel Hell found that in human–human relationships the depressed person increasingly dissociates him/herself, the more (s)he feels misunderstood by the partner, who often attempts to help. The cat accepts the level of 'interactivity' the (depressed) owner wants to have and is present when the owner desires that contact without forcing itself on the human partner....To some degree, cats can be more pleasant partners for depressed people than humans."[5] I wonder if Turner was smiling, perhaps grimly, when he wrote this.

Other scientists contend that the studies showing positive health effects were methodologically flawed, and the benefits are illusory, except perhaps some improvement in self-reported happiness.[6] (One study found that the only measurable difference between pet owners and non-owners was that the owners were fatter.[7]) Regardless of measurable factors, as Richard Mayon-White of Oxford University writes, "Good health is more than the absence of disease."[8]

What I consider to be definitive, a meta-study (that is, a study of studies) by the Human Animal Bond Research Initiative Foundation titled "The Health Care Cost Savings of Pet Ownership" not only affirmed the findings of "a growing body of academic and professional research" showing "the health benefits of pet ownership" but also showed dramatically lower "total health care spending."[9]

Only eleven of Karsh's seventeen adopters stuck with their cats. What had gone wrong in the other six cases? Could they just have been bad matches? Might there be a better way, she wondered, to choose a kitten than "I like the look of that one"? Karsh and her colleague Carmen Burket, a clinical psychologist, developed a program they called Companion Cat, which relied on structured interviews with prospective adopters. The interviewer would describe a number of aspects of cat reality—appearance, yes, but also various aspects of personality and temperament—and then the interviewee would be asked to express a preference in each category. Level of "activity" would be an important one: Did they prefer a cat that was a sleepy, dreamy couch potato, or a spitfire, or an up-all-nighter, or a princess? Some people might be bored to death by a lethargic cat. A highly active cat might drive another person nuts.

The next part of the program, then, was to create a profile of the adopter's personality. "People who lack self-confidence," writes Karsh, "find it difficult, even impossible, to relate to a shy or timid cat….The cat's inclination to hide triggers feelings of rejection in the insecure person….Nurturant people, however, provide good homes for timid or even fearful cats." She considers "as an ideal person-to-cat match, placing aggressively friendly cats (those who come up and repeatedly initiate cat–person interactions)"—the kind of cat her early-handling experiments were producing quite a few of—"as therapy for mild to moderately depressed persons."

Karsh and Burket collided with several intractable adoption problems. The most common occurred when people's cats had died only recently, and the grieving owners hadn't found their way through their grief: Often they wanted a cat just like the one they had lost. Which of course was impossible. Over and over they were disappointed, and the poor unknowing substitute would be returned, or, worse, put down.

The second problem would be soap-operatic if it weren't so sad. An older person, usually but not always a woman, is living alone. Her adult son or daughter, or more than one of her children conspiring, decide that Mom's loneliness would be wonderfully assuaged by a cat. Maybe they've read how some elderly folks' dark, dull lives have been pepped up and brightened by a kitty. Except that Mom doesn't want one. Eventually she gives up and they go to Karsh's lab, where they've heard the kittens are so sweet. Mom can't make up her mind which one she wants, so one of the kids picks one. A few weeks later Mom calls and says the cat doesn't like her, it's got to go. In fact what has happened is that she has made zero effort to love the cat and the cat has responded in kind.

The third difficulty was the plain old mismatch—the timid cat and the depressive, the manic cat and the already nervous wreck.

Over time, Karsh got better at matchmaking, and she was particularly happy with how things turned out for the adopters who were lonely, elderly, handicapped, or depressed. Single parents found something new to share with their kids, something all could love equally without a burden of competition. One older woman's systolic blood pressure fell from 175 to 125. Another, ninety years old, could no longer hear her phone or doorbell ring; her new cat learned to summon her when either rang. "If you talk to some elderly people, all they have to tell you about is their infirmities," Karsh said. "When they get a cat, they have cat stories to tell."[10]

A study by Adnan Qureshi and colleagues of the University of Minnesota looked at the occurrence of fatal heart attacks and strokes in a sample of more than four thousand people. After adjusting for a number of other predictive factors—gender, ethnicity, blood pressure, smoking, diabetes, and obesity among them—the study showed that cat owners had a significantly smaller risk of deadly heart disease than people who had never had a cat.[11]

The cats that the six original adopters in Karsh's study gave up on were lucky, because it was understood that if the adoptions didn't work out she would take them back, and in time she found good, permanent homes for all six rejected cats. The more typical outcome of a failed adoption, however, is that the cat ends up euthanized in a shelter. The shelter is brimming with kittens, younger and cuter than the rejectee, and they're the ones chosen and happily carried home.

Karsh points out that nearly all kittens are "acceptably friendly.... However, if they have not been handled during the socialization period…they will typically grow up to be much more aloof. Also, unless one observes many kittens, it is difficult to tell an overly active kitten from an ordinarily active one.…Therefore, if a kitten is selected impulsively"—almost always, to Karsh's eternal frustration, by appearance alone—"it may very well mature into a cat that differs from the adopter's expectations." And end up just another older cat back in the shelter.

Most shelters have more cats than will ever be adopted. The cutest kittens win. The others languish until the staff and volunteers face the agony of killing them. In no-kill shelters, of which there are increasingly many, finding a home for the cats no one wants can be a struggle—phone call after phone call to other shelters, municipal agencies, farmers' associations, until somebody has an open adoption slot or a barn in need of a mouser.

It is a baffling fact that a good many people don't care. A good many people—mostly men, it seems, God knows why—just hate cats. Cats were worshiped in Egypt, loved in ancient Rome, and loved well into the early years of Christianity in Europe. Then in the Middle Ages cats came to be associated with witchcraft and the plagues, and were widely reviled, tortured, burned. Now they have come back into the circle of Western people's love, but incompletely. Black cats still spell bad luck to many (and are, in fact, adopted less often). Some people sense a vague sort of evil in all cats. Some people fear them. Karsh writes that on several occasions, groups of workmen were afraid to enter her cat colony rooms, with ten or fifteen cats roaming around loose, until somebody locked them up.

Some people hate cats for their apparent disposition not to communicate more clearly (a false impression, as we will see). Many believe them incapable of affection, unworthy of existence. If the dog hates 'em, they think, he must know something.

A great many more people love cats. They are by far the most popular animal companion in the world. There are ninety-six million pet cats in the United States, ten million in Canada, ten million in the United Kingdom, three million in Australia, a million and a half in New Zealand, fifty-three million in China, ten million in France, nine million in Italy, eight million in Germany, seven million in Japan. These are owned cats; we'll deal with feral cats later. Some small number of these are barn cats and such, unneutered tom roamers, kitties so shadow-shy they won't let you touch them—but the vast majority are our familiar, curious, weird, intermittently affectionate, unpredictable, dinner-loving house cats. Almost all of them are loved, albeit sometimes with a dose of frustration.

There are more cats than dogs in the United States, but there are more households with dogs. That's because households with more than one cat are fairly common and multi-dog households much less so. But there are more pet fish than either.[12] Less beloved, safe to say.

Victoria Voith, a professor of animal behavior at the Western University of Health Sciences, who is a veterinarian with a master's degree in psychology, interviewed 872 cat owners who brought their cats to the University of Pennsylvania's veterinary hospital in 1981 and 1982. Ninety-nine percent of them considered their cats to be a family member. (The Humane Society of the United States, however, publishes a very different number: Citing the *American Veterinary Medical Association Sourcebook* of 2012, they state that 56.1 percent of American cat owners consider their cats to be family members.[13]) In Voith's study, 97 percent of the cat owners talked to their cats at least once a day. Ninety-one percent believed their cats were attuned to their moods. Eighty-nine percent of the cats slept on their owners' beds.

Nearly half these people reported that their cats had bad-conduct issues of some sort—mainly litter box problems and destructive behavior—but nobody was considering getting rid of their cat. This intrigued Voith, and she then designed another study "to learn why owners keep pets despite serious behavior problems." She focused on people who had brought their cats to the hospital specifically for treatment of such problems. She asked thirty-eight of them why they'd hung on to their troublesome cats for so long. Fifty-five percent answered with something like a shocked, "Why, I love him!" or "You wouldn't get rid of a child because he had a behavior problem, would you?" Obviously, no matter how much trouble they were, these cats were full-fledged members of their families.

"If an owner decides not to keep a pet because of a behavior problem," Voith continued, "the decision is usually reached with considerable sorrow and regret. [The owners] often acknowledge a paradox. They recognize their attachment to the pet and, at the same time, acknowledge that the animal is a source of inconvenience, a financial expense, social concern, or even a physical threat. They will say, 'I know he's just a dog, but I feel about him as though he were my child.'" In fact, Voith concluded, "Many of the behavior patterns between people and pets are very similar to those that occur between a person and a child."[14]

"Like children," writes Eileen Karsh, "your cat (or dog) is dependent upon you for survival and, in this respect, is a perpetual child. Your cat appears to miss you when you are away and be happy to see you when you return. And your cat relates to you as an individual in a non-judgmental way, independent of your professional, social, or financial success or failure."

⌒

Finding a cat who has been raised according to Karsh and Turner's findings isn't easy. The best shelters' fostering programs do come close, and are surely your best bet. But cats come into our lives in many other ways than by rationally decided and planned adoption of a gently fostered kitten of a good mother at the age of at least three months. Augusta was dumped in the snow and survived a night skedaddle down the driveway. Sometimes your mother dies and leaves you her cats, all three old, one cranky, one sick, and one who hides from you for the first six months. Sometimes a cat shows up at your door and won't stop complaining till you feed him, and eventually you let him inside. Sometimes the cat is part of the girlfriend package and you and your dog both have to learn not to hate cats, even while this one has peed on two of your sweaters—which you're going to have to learn not to leave on the floor. You might see a cardboard box of mewling six-week-old orphans in the lap of a young woman on the sidewalk in the evident grip of despair, probably homeless, possibly a junkie, and on the box are shakily lettered the two words FREE KITTENS.

Whichever cat or cats you end up with, however it happens, you will be in for some surprises. If you've had cats already, you know. "Cats vary as much within their species as people do within theirs," wrote the late Muriel Beadle in her well-known book *The Cat*. "Some cats are so dull-witted that never in their lives do they learn that a door which is ajar can be nudged further open. Other cats can open shut doors and get the lids off garbage cans as easily as if they had two hands with opposable thumbs. Some cats are born with unshakably equable dispositions and others are so high-strung that they go all to pieces when a doorbell rings."

In "Individuality in the Domestic Cat"—published in the same collection as Karsh and Turner's study—Michael Mendl and Robert Harcourt search for possible evolutionary explanations for the otherwise inexplicable eccentricities of the species. "The process of domestication is likely to have relaxed the constraints imposed upon the cat's behavior by natural selection," they write—speculatively—"and to have allowed it to express more variation in behavior than is seen in non-domesticated species. The cushioned existence of many domestic cats alleviates the constant need for behavior to be continually directed toward catching prey, mating, avoiding predators, and so on." Which leaves them plenty of time for jumping out of hiding at you, racing mindlessly around the room, playing (or refusing to play) with their wide selection of cat toys, lying on whatever you're trying to read, unspooling an entire roll of toilet paper, or… *playing with bears in the dark.* Individuality, thy name is cat. Just a few examples:

> Charlie, my angel boy, used to get his toy mice right after I'd gone to bed. He would meow that really long and sad meow, and pick them up one by one bringing them up on the bed and laying them down beside me. Then he would lie down on the other side of me.
>
> My cats get lost in the house. They'll go downstairs in the night to eat or use the litter box and then stand at the bottom of the stairs and cry. You have to call out to them so they can follow the sound of your voice back to you. One's lived in this house for six years, the other almost three.
>
> Our cat, Maverick, jumps up the sides of the walls separating our family room and dining room. He jumps almost four

feet, puts his "arms" on each side of the wall and slides down. He's done it since he was a kitten.[15]

Elizabeth and I had a friend with a long history of training animals, including cats—she was the one who told us how to teach Augusta to fetch. Shamed by our failure to carry that through, we tried other tricks that our friend, again, assured us we could teach her easy-peasy. You have a bag of little crunchy cat treats, and each time Augusta does something that remotely resembles the trick, you praise her, pet her, and give her the treat, which she loved. We couldn't get anywhere.

We concluded that this was a case of *individual variation*. Some cats were smart; some were not. Walter and Penny, the barn cats, obviously had all kinds of things figured out, for example their calm fencepost-sitting when coyotes came to call. It looked as though we had to face the fact that Augusta was just plain dumb. She was beautiful, though! So sleek, so smooth, so black. Everybody said how pretty she was. People do come to characterize their cats on a combination of available evidence and a certain amount of fantasy. We didn't know, we just liked the idea of her being the black cat equivalent of a dumb blonde, and one of her affectionate nicknames became Dummy. Also Stoopie.

The interesting question does arise of how, if you're a serious researcher, you measure things like individual intellectual variation in cats. They don't take tests very well. A fine description of the challenge comes from Alex Thornton and Dieter Lukas, a psychologist and a zoologist, respectively, at Cambridge University, in their introduction to a paper titled "Individual Variation in Cognitive Performance: Development and Evolutionary Perspectives":

Imagine a team of alien scientists visiting London during the summer of 2012, selecting a random sample of twenty humans and conducting experiments to test theories of human evolution. Some trials involve swimming, and most subjects perform rather poorly. However, one subject happens to be Michael Phelps, the Olympic record holder. Based on Phelps's performance, the aliens conclude that humans have an astounding capacity for high-speed movement through water, underpinned by physiological and behavioral adaptations including efficient conversion of stored carbohydrates to sugars and fine-scale motor control for efficient propulsion. From this, they argue in favor of the aquatic ape hypothesis, which postulates that ancestral humans were under strong selection for an aquatic existence.

The paper—essentially a review of a slew of studies of cognition in many species—concludes that most of the work done up to the time of its publication, in 2012, was methodologically flawed because it didn't adequately consider the wide differences among individuals. They suggested that some variation could represent "adaptive plasticity in response to local conditions," and said also that "we must ask whether individual differences in cognitive traits are heritable and whether they have consequences for reproductive fitness."

Perhaps, then, Augusta was dumb because she didn't *need* to be smart. Reproductive fitness wasn't an issue, because she was spayed, but all the wild and crazy minds, ideas, and imaginations of other cats, some number of whom were certainly fertile, could make for a dandy study of the question of their reproductive fitness—just as soon as somebody could figure out how to measure—or even characterize in comprehensible terms—the nature of their variations. Then tabulate same. Bon voyage!

ℐ

Temple University closed Eileen Karsh's lab in 1990. She then moved to the suburban Ambler campus, which had a population of feral cats. She worked with her students trapping them, feeding them, and gradually making them more or less tractable enough to be adopted. She continued teaching until her retirement in 1997, but she no longer conducted formal research. Her epochal study, "The Human–Cat Relationship," of 1986, when she was only fifty-four years old, was the last scientific paper she ever published.

∽

Was Augusta happy? We know now that her environment provided the right ingredients. That was luck—living on a ranch, far from the road, no traffic on it anyway, nice dogs, nice horses, nice cats, no nasty predators so far, lots of fascinating habitat to explore. She had owners who loved her and who tried, in their own dumb ways, to make her happy. But I believe that another of her individual traits was an innate happiness, a capacity for joy, which found its incarnation—never for a moment the same yet always the same—in the sun and the grass and the space and the wind of her home in Montana, its plants and animals, water and soil, its whispers and its fragrances. She was of that place, that ecosystem. Its gift to her—and, through her, to us—was her capacity for absorption, engagement, all-attention. No, those are too small. Call it delight.

We recognized it fully only in retrospect, by its absence. When life in my ranch partnership came to be too expensive, I had to sell my share, and we moved to San Francisco. We feared that as Augusta's range shrank from a boundlessly fascinating universe to the narrow confines of apartment life, it would dull her soul pitifully. Then we found a sunny flat with an exterior staircase and a back yard separated from others only by cat-climbable fences. Perhaps this could prove a new frontier—it had texture, complexity, challenge. She was just turning two years old, and brimming with vitality, as well as that feline essence, curiosity. She began to teach us her new needs—new games, new races and chases. She charged up and down the back stairs, tightrope-tiptoed the banister top rail, prowled the scruffy, untended bushes at the bottom of the yard. Surely this was engagement? At first we were so busy, and so jangled ourselves by the shock of our own abrupt and entire change of habitat, that we did not see the faint dimming in her eyes.

Her new toys and adventures seemed satisfactory. She purred at Elizabeth's feet in bed as always before. But as the picture formed in our memory of her life in Montana we saw there the glossy ripple of light in her fur, the acuteness of attention in her ears and whiskers, a lightness of movement that we had not been seeing here in the city. We began to notice that now there were noises that startled her, sirens, airplanes, car horns, shouts in the street, the whine of saws and whamming of hammers from the building next door, sounds that we, as longtime urban dwellers, easily absorbed and more or less forgot. For Augusta, perhaps even the anticipation of them induced anxiety.

Then my father came for a visit, and some rough barking laugh of his, or some clumsy attempt to pet her like a dog, sent her pipping out the back door and gone, into the morning, into the afternoon. We took the old man to the airport. When we came back Augusta was still gone. Into the night. We called and called. She always came, but she didn't come.

Late, late that night, when the traffic noise had died, from the building next door, which was under renovation and empty, Elizabeth heard a weak, miserable mew. We called to her. Come home, little kitty, oh, dummy, come home! *Please. Mew.* We climbed into the adjoining back yard, found a ladder, squeezed through a window, worked our way through three flats littered with construction debris, sweeping our flashlights, calling as calmly as we could (not very). From somewhere, we couldn't tell where, *mew*. There was still a low, unfinished attic to look in, and in the highest corner of it, wedged into the tiniest crevice, there was Augusta, frozen in…misery? terror? In any case no longer even speaking. Elizabeth had to pry her out, and she was ajitter for days.

We bought a house in the street behind, a narrow old Victorian with a little back yard—a place where Augusta could sniff, climb, roam, and yet be protected from traffic, for the entire block was enclosed by buildings shoulder to shoulder. We would see how that would be, but meanwhile summer was coming on and a dear friend was lending us her place in Montana, near my old ranch.

It was near indeed—the main Boulder River—but the main was a different world from our old West Fork of the Boulder. The West Boulder was rocky, rough, a hurrying mountain river. The water here was wide and slow, and it soaked into the meadowed plain it meandered through, an ancient lake bed, giving rise to quiet little creeklets and springs and a magnificent old-growth cottonwood forest. Augusta took one look at that cool, shadowy woodland, raised her nose, breathed it in, and said, Yes, this is mine. What had been absent in San Francisco confirmed its existence here so unmistakably that it might as well have been a notarized declaration: This is who I am, what I am.

She went to the cottonwoods first thing each morning with such regularity that we called it her office. Sometimes she worked late and had to be called insistently home, but she would always come, plunging through the yellow-green grass, which grew tall in those spongy riparian meadows, her little black head bobbing up-up-up, boink-boink-boink, porpoising, eyes on us at the top of each arc. We'd be laughing and urging her on, Come on, Boopus, come on!

Once, a pal came to visit with his excessively friendly Irish setter, the kind who greets a cat with a genial, loud-arfing charge, no harm meant, just want to say hi, doncha know—not Augusta's notion of gentility. Met thus halfway home from the office, she fled back into the cottonwoods, and then came a sudden late-afternoon deluge. We knew that no amount of calling at that point would dislodge her, that a long pause for dinner and the confinement of the sweet doggy in our friend's pickup were the only hope, but no dice.

We set out in the pouring dark, stumbling over downed branches, splooching into mudholes, and when at last we found her she was still invisible but plenty audible, about twelve feet up, in a deep mossy bowl that had formed where the huge cottonwood tree had broken off at that height long ago. We staggered back home, searched out a ladder, and then had to find our way back to the tree, no mean trick. Augusta was so soaked her teeth were chattering.

This version of the rescued Augusta was quite different from the one in the San Francisco attic. As we toweled her roughly back into shape she was purring and pliant, and after that she didn't stop grooming herself till, finally dry, she fell asleep and slept all night, soft all over. First thing in the morning, sun bright, job to do (whatever it was), back she went to the woodland, all delight.

Chapter Three

Thinking? Talking?

What was she saying? We were sure it was something, and from the look on Augusta's face you could tell she was distressed that we weren't getting it. She tried saying it again, in several different ways. It wasn't "out," it wasn't "food," it wasn't "pick me up," or any of the usual requests, nor did it sound like any of them.

How much can you know of what your cat knows? Most cats' human companions take one of two approaches: Either they project their own imaginings onto kitty—ranging from serious wish fulfillment to preposterous YouTube fantasy—or they indifferently reject the question as unanswerable, often in fact valuing their cats' opacity. I would watch Augusta for hours, lost in wonder, learning almost nothing. Charles Darwin devoted an entire book to *The Expression of the Emotions in Man and Animals,* finding by studious observation that at least some animals—the more expressive ones, like dogs and chimps—shared a wide range of emotional understanding with humans, and we've now seen that people and cats have a deep-seated neural substrate in common, which recognizes each other's vocal expression of basic emotions. (You will also recall that that recognition may be only unconscious in humans.) Other research, some of it quite recent and not widely known, has begun to reveal the astonishing scope and subtlety of what cats know and how they express it to us.

Cats' communication with other cats is part of what we need to understand, but far from all. A lot of cat-to-cat information exchange is body language, and people watching cats over time have come to understand it pretty well. Cats use most of the same postures, expressions, and gestures to express the same things to us. Most cat books describe the basics of feline body language—the flattened ears and thrashing tail of the angry cat, the hair-on-end and hiss of the frightened one, the scamper and horseshoe-tail of let's-play, and so on—but this chapter's exploration of vocal expression will show that there are more nuance and variety in cat language than most people have ever imagined possible.

Cats' communication also employs their genius for olfactory discrimination, which we get only the occasional and unpleasant merest fraction of, alas, as when kitty sprays her pungent, invisible message where we don't want it. They try to communicate with us in scent: The simple chin-rub and head-bonk are gestures of familiar affection that also have a significant pheromonal component, if only we could smell it. Equally invisible to us—and unintelligible because unsmellable—are sebaceous glands not only on the friendly chin and noggin-top but also along the jawline, along the lips, at the base of the tail, and just inside the ears. Each secretion is host to a complex of bacteria and other microorganisms that produce a variety of olfactory signals. Depending on your cat's preference, or tolerance, a gentle or strong rub on those sites can calm and please your cat. There are several other gland sites that you'll have to use careful judgment about: the flanks (some cats can't stand to be touched there and will scratch you in a flash, instinctively), between the toe pads (ditto), and just around the anus (even if you can stand the thought, be very careful there). A patient and attentive ear and eye will be rewarded.

Thinking? Talking?

Neutering greatly reduces urine spraying, but for owners of intact cats, especially males, it's obviously an important concern. Cats are territorial, they just are—especially unneutered tomcats—and establishing and maintaining territorial boundaries are ineradicably inscribed in their DNA. They also need to know all about every other cat boundary that touches on their own. This is as essential to a cat as food. All cats, therefore, fixed or not, need to read freshly spray-marked surfaces. When cats first encounter the litter boxes of other cats, they pay close attention to whatever information is encoded there, and what that may be, beyond proof of ownership, we really don't know. Someday, perhaps, researchers will be able to read the rich and various scent code of cats, but so far science has yet to crack it. We know it's way more than territory marking and sexual signals, but a translation of the cat's vocabulary of scent into anything meaningful to the human mind remains an unexplored opportunity.

Researchers have, however, spent years on the mystery of the purr—a form of expression that begins between kitten and mother, in mutual soothing, but which then essentially dies out in adult inter-cat communication. After that private maternal phase, the purr becomes exclusively a cat's expression of something to a person: They do not purr to one another. The general belief is that it means the cat is happy, which, in general, isn't wrong. Some people say it means the cat loves you. You may also have noticed that purring increases toward mealtimes. Some behaviorists believe they've teased out a certain occasional scamming in the purposeful purrer. It is also the case, however, that cats often purr when they're sick, or injured, or dying.

It took science a long, long time and a lot of busted hypotheses to figure out how cats produce the remarkable suite of sounds we group together under the rubric of purring. In fact, it's not entirely figured out yet. In the breakthrough (and still gold standard) paper

"How Cats Purr," published in 1991, Dawn E. Frazer Sissom and colleagues write, "Despite many years of observation, our understanding of the mechanism of cats' purr is still incomplete. Questions remain concerning which organs are involved and what processes produce the sound and vibration."[1] The Frazer Sissom study used high-sensitivity oscilloscopes and microphones to record purring not only at the mouth and nose, where it's loudest, but all around the body surface. In fact, if you think about it, we commonly perceive purring as much as a tactile experience as an auditory one.

The auditory aspect is extremely complex. Please do try to grok these details—the whole thing is weird. The fundamental frequency of the purr in all domestic cats is about twenty-five hertz (cycles per second, abbreviated as Hz). With only narrow variation, that tone is the *same* in cats of *both* sexes, *all* sizes, and *any* age, from kittenhood to senior citizenship. Twenty-five hertz is lower than the lowest key on an eighty-eight-key piano. It's barely above the average low limit of human hearing—twenty Hz—and for that reason quite a few people can't hear the basic tone at all. What anybody with ordinary hearing hears are the *overtones* produced by the fundamental frequency. It is the "chordal" nature of the overtones that make the cat's purr such a rich and complicated sensation. Usually, both the sound and the palpable vibrations hardly vary between in-breath and out-breath, and the pause between in- and out- is barely perceptible. Because the fundamental frequency is so low, it is audible only within a couple of feet. This is truly intimate communication.

The mechanics aren't simple, or they would have been explained long ago. Studies prior to Frazer Sissom's had gone down many blind alleys. There were all sorts of theories about the vocal cords, all wrong. (It should be noted here that what most everybody calls the vocal cords are, in fact, vocal *folds.* That is the term scientists use.) Electrodes probing cats' heads located several areas of the brain that elicited purring when poked—proving at least that the central nervous system was involved, but decidedly not establishing the "purring center" in the brain that had been hypothesized. Electromyographic instrumentation of cats' larynx and diaphragm muscles showed activation at the fundamental purring frequency—but then cutting the poor lab cat's laryngeal sensory nerve did nothing to stop the purring. Other experiments claimed to show involvement of the intercostal muscles (those between the ribs). So the purr was originating, apparently, at several loci in the brain, and nobody seems to have been able to find even where the purr was being physically produced.

Finally the Frazer Sissom study nailed it—the basic mechanism, anyway. The intercostals had nothing to do with it. When your cat purrs, his larynx acts as an extremely fast-opening and fast-closing gate; the instantaneous onrush of breath and the equally instantaneous intake force the vocal folds suddenly open just to the point at which they produce, as Frazer Sissom et al., write, "a sound very rich in harmonics (i.e., the fundamental tone). The vocal tract filters this sound and conducts it to be radiated from the mouth and nose." It's important to realize that even though the vocal folds are producing the sound, they're not doing it in the same way as when we talk or the cat meows—those are much higher frequencies, and involve varying the tension of the vocal folds and, hence, their length; that's how you get the great variety of voice that both people and cats are capable of. In purring, the vocal folds don't stretch; they just open and close really fast. And as for the humming you feel under your hand, "The surface vibrations are caused by the same pressure difference across the larynx as is the sound. Pressure changes propagate as sound waves from the trachea to the surface of the lung at speeds approaching three hundred m/s (meters per second)." You will notice that you feel the purr most distinctly just where your kitty's lungs are.

Is this not amazing?

Given that purring is intended for human understanding alone, it makes sense that variations will have evolved for specific expressions. Any cat owner knows that one of the beast's favorite themes is, "What about my dinner?" The seamless transition from the purr of pure contentment to the barely audible, just slightly irritating extra layer of urgency is a fine example of such evolution. It's physiologically possible in the first place because the purring cat's vocal folds can continue doing their breath-gating job while at the same time stretching to vibrate up in the voice frequencies. Cats have also evolved a particular "isolation cry" ("I'm lost!") with a remarkable acoustical resemblance to the distress cry of a human infant, to which humans are instinctively responsive. And now wouldn't you know, the acoustic profile of the subtly added vocal note—"the cry embedded within the purr," as it was called in the study that first identified it[2]—falls right in the same frequency range as a human baby wailing. "The inclusion of this high frequency component within the purr," the study's authors write, "could serve as a subtle means of exploitation, tapping into an inherent mammalian sensitivity to such cries and also possibly rendering the call less harmonic and thus more difficult to habituate to." Exploitation? By a cat? Yep.

And this little trick of emotional blackmail tunes to neuroacoustic signals buried deep in your brain a long, long time ago. Anybody still want to say cats are dumb?

𝒯

That brings us neatly to the bigger question of what cats actually know, a subset of which is what they say. Modern science has been singularly unhelpful here. To a lot of biologists, the question of animal intelligence is kryptonite—touch it and your own intelligence will be melted by the mockery of your peers. A brave minority, nevertheless, have devised experiments in which birds, for example, have performed remarkable feats of what can only be called brilliance. The obstacle to examining animal intelligence has always been not its absence or low quality but the difficulty of devising situations and tests congenial to the nature of the animal under consideration—and obtaining the cooperation of a subject who is quite likely not to see the point of the constraint, obedience, occasionally pain, and, often, from the subject's point of view, absurdity involved. Some species, of course, are more cooperative than others. Dogs come to mind.

Some animals seem to be just *too* smart to play along. Horses are for the most part wondrously docile, and can be trained to do the damnedest things on command; but anyone who knows them well can tell you they'll play wicked tricks on you too—and when you least expect it. Some of the smaller cetaceans (dolphins, porpoises, and whales) have the same qualities, but they seem also to keep the greater part of what they know entirely to themselves. Or else they're trying to tell us things and we're just too dumb to have learned their language.

And so modern science, for the most part, has had to forego translation of animal language and content itself with the observation and interpretation of behavior. When scientists manage to see that behavior on the animals' own terms, it can, as Stephen Budiansky writes, be nothing short of miraculous:

Evolution, learning, the very wiring of animals' brains and sense organs, adapt them to the cognitive demands of their physical and social environments in ways that at times put us to shame, with our reliance on consciousness and language and reason to see us through. We can study books about cryptic moths and make long lists of their distinguishing characteristics and actively try to commit them to memory and in time can do as well as every blue jay in the world can. We might with enough mnemonic devices or other memory tricks find caches the way Clark's nutcrackers can. We can never follow a trail the way a dog can, or find a rotting carcass the way a turkey vulture can, or navigate our way home the way a pigeon can, or locate a tiny rapidly moving target the way a bat can, or calculate the distance of a Carolina wren from the tonal shifts in its song the way a Carolina wren can.[3]

As we'll see in chapter 5, cats have their own system of knowledge—an immense amount to tell the scrupulous analyst of their behavior. They also insist on *talking* to us. And, almost always, only to us. The only talking that grown-up cats do among themselves has to do with sex, and those vocalizations are so closely tied to specific behavior that they have proved relatively easy to interpret. Making sense of the sounds that come out of cats' mouths in the company of people is not easy at all.

Those sounds—the myriad variations of *meow, aow, mmph,* and the rest—are made only for the sake of saying particular things to particular people, and among all the felids of the world, only domestic cats try to talk to people. A Cornell University biologist named Nicholas Nicastro was "interested in learning how humans have shaped cat vocal behavior by artificial selection, and how cats have evolved to exploit pre-existing human perceptual tendencies." (There's that word *exploit* again.) "Seven thousand years ago," he says, "when we think the ancestors of our domesticated cats began wandering into Egyptian granaries and offering to trade rodent-control services for shelter, it was probably the pleasant-sounding cats that were selected and accepted into human society." Just to see if our cats' ancestors showed any of the same proclivity for making nice, attractive noises to people, Nicastro visited some African wildcats at the South African national zoo. "Those cats sounded permanently angry," he wrote. "If they were looking for affection, they weren't expressing themselves very well."[4]

Not only have cats developed a language for humans in general, they often customize it for the individual. Most cat owners— 97 percent, in Victoria Voith's study—talk to their cats, and over time the language, or dialect, of which they share an understanding evolves its own particular shape, elements, rhythm, and melody. There are constants as well, the biggest reason being that cats are capable of producing only a limited range of basic sounds. Within that range, nonetheless, is great diversity. All this complexity—the individual variation between particular people and particular cats, the bewildering variety of subtle differences of expression—has had the evident result that modern science won't touch cat talk with a ten-foot pole.

That is, scientists won't attempt trying to *translate* it. There is, however, a good deal of fine-grained physical analysis going on. Susanne Schötz, a professor of phonetics at Lund University in Sweden, has done remarkable work in characterizing the acoustic properties of nearly every sound ever to come out of a cat's mouth. Just one of her many papers—don't laugh, this is serious science— "A Phonetic Pilot Study of Chirp, Chatter, Tweet, and Tweedle in Three Domestic Cats"[5]—establishes a virtual taxonomy for the vocal output of three cats looking at birds through a window. What Schötz's three cats had to say comprised not only the four "words" in her title but also chirr, chitter, chirrup, peep, pipe, twitter, warble, and quaver. Those were ultimately grouped together as either chirps, chatters, or tweedles. (Okay, go ahead and laugh.) A computer program calculated the mean duration, the minimum basic frequency, the maximum basic frequency, and the mean basic frequency of each. The paper also classifies the voice (voiced or unvoiced), the pitch, the loudness, the rate, the "modulation/reiteration" (for example, "rapid series or rattling"), and "other descriptions or comments" (e.g., "clicking sounds used to urge on a horse" and "titter, nervous giggle") for six basic vocalizations. As for what they may have *meant*? Professor Schötz doesn't address that.

In another paper, Schötz and her colleague Joost van de Weijer did consider not what cats were trying to say, but what people *thought* they might be saying. The focus of "Human Perception of Intonation in Domestic Cat Meows"[6] was pretty narrow: cats meowing at feeding time and cats meowing while waiting at the veterinarian's office. They found that the hungry cats' meows went up in tone at the end, and the anxious cats' meows went down. They asked people familiar with cats to interpret recordings of those meows, and people who weren't familiar with cats. The people who were familiar with cats were more accurate in their perceptions. "Taken together," wrote Schötz and van de Weijer, "these results suggest that cats may use different intonation patterns in their vocal interaction with humans, and that humans are able to identify the vocalizations based on intonation."

And that's about the state of the art in contemporary research on cat talk. Luckily, a work of genius produced many years prior to the Swedish studies did the kind of comprehensive analysis that seems to be missing from the current scientific literature.

"Vocalizing in the House Cat: A Phonetic and Functional Study," by Mildred Moelk, was published in the *American Journal of Psychology* in 1944. It is a masterpiece not only of observation and interpretation but also of wit, style, and grace.

Moelk undertook her work with simplicity: She studied her own cats—very closely. Once she had arrived at her categorizations of their speech and her conclusions about its intentions, she checked them far and wide, both observing other cats and also studying the work of observers from other cultures, other languages, and other times. But what seems most significant, and most beautiful, about her work, is the quality of attention she brought to it, particularly in her exquisite contemplation of a single cat—"a female given to much vocalizing…already in her tenth year when the study began, [who] had long before achieved an equilibrium of acquaintance and habit between herself and her environment…[and whose] vocalizing patterns may therefore be regarded as firmly established.

"Of the sixteen phonetic patterns to be distinguished for this mature cat," she continued, "all but three have been observed in other subjects as well. Two of these ('bewilderment' and 'refusal') are of minor importance, while the third ('acknowledgment') is one which could be expected only from a cat very much at ease in the presence of the auditor."

Very much at ease. That state of mind is what made Moelk's observations possible, and Eileen Karsh's too. And it is the *absence* of ease that continues to contaminate a great deal of research on animal behavior. When your experimental subject cat is nervous or afraid or angry or freaked out—the usual situation of a cat being *experimented on* in a *laboratory*—you might as well just forget the experiment and go home.

Here is a supreme example of an experiment that should have been forgotten in advance. The researchers proposed to use a modified version of the Ainsworth Strange Situation Test to evaluate "the claim that the cat–owner bond is typically a secure attachment."[7] The deal was basically to see if cats behaved measurably differently toward their owners and toward strangers. What made the experiment insane was the setting. A few details will suffice:

Two similar, plain rooms were used for the study, in order to ensure an equivalent strange physical environment.…

Windows in both rooms were covered to avoid any visual distraction....The rooms were divided by strips of white tape into four zones: i) region around owner's chair, ii) region around stranger's chair, iii) door, iv) play area.... The test rooms and all equipment were thoroughly cleaned with an enzymatic cleaner (Urine-Off for Cats and Kittens) before and after each test....

Can you imagine? These people thought they could get meaningful behavior from cats in a setting like that? "Two cat subjects... hid during an entire experimental testing period and were, therefore, removed from the data analysis." You don't say!

Mildred Moelk's cats were at home and "very much at ease." She didn't manipulate them in any way. She didn't cover the windows or stink up the room with an enzymatic cleaner. She just listened, and watched. She also had a thorough command of linguistics and phonetics, so she knew *how* to listen.

Moelk's report of her study begins by reminding us that cats and people produce vocal sounds quite differently. We talk only on the out-breath. Cat speech uses both in-breath and out-. Cat talk doesn't use the tips of their tongues at all. We move our jaws, lips, tongues, and other mouth parts to form words; cats mostly change the tension in their throats. She also warns us in advance that she's going to use non-scientific language like "bright," "dull," and "eager." And we learn up front that she is going to have to use symbols from the International Phonetic Alphabet to represent particular sounds— symbols that aren't at all easy to remember unless you've studied phonetics (examples: *ŋ* for the sound at the end of the word *song*; *œ* as in the French word *oeuf;* ~ to indicate "nasalization," and so on)—sounds that can't be otherwise described in print but which she had to learn to listen for in order to discriminate among the "words" her cats spoke.

Thinking? Talking?

Moelk builds her system of understanding on a simple foundation of three main classes: "1) sounds produced while the mouth is kept closed, the murmurs, 2) sounds produced while the mouth is opened and then gradually closed, resulting in a fixed-vowel pattern, basically *a:ou*, and 3) sounds produced while the mouth is held tensely open in one position." She goes on to explain: "The differences between these three classes are at bottom differences in degree of intensity of effort. More energy is required to say 'What?' than to murmur 'Mhn?' and still more to snap or thunder 'What!' than to make the calm inquiry." Already we have set sail on the sea of real language.

The intensity and precision of Moelk's ability to listen combine with her understanding of her cats' intentions to yield accurate (and beautifully expressed) interpretations of their speech. An example, among the sounds produced in her first category of vocalizations, "murmur patterns"—in which the cat's mouth remains closed and the sound comes only through the nose—is "the 'Level *'mhrn* of Request or Greeting': In structure the *'mhrn* resembles an isolated and expanded form of the inhaled stroke of purring. The initial *m* indicates its greater body. The quality of the *'mhrn*, like that of the purr, involves the presence and brightness of the *r* element. Unlike the *'hrn* of purring, however, *'mhrn* is not composed of a series of separately audible vibrations...but is prolonged by means of repeated new intakes of breath before release through exhalation, represented by repeating the *'hr, mhr'hr'hrn*. Murmurs of one, two, or three such rolls are most common, although murmuring may be protracted even through the length of time it takes a cat to run up a flight of stairs, continuing a single *mhr'hr'hr'hrn* all the way."

Can you not absolutely hear that cat running up the stairs?

One after another, Mildred Moelk listens deep into cat words, spells them out sound by sound, and names them: the Call; the *'mhrŋ* of Acknowledgment or Confirmation; the Demand. Within each can be many variations, each offering its own fine distinctions of meaning: "As the demands become more and more vociferous, increasing stress is placed upon the initial vowel while the murmur-element contracts and intensifies.... A relaxing of stress upon the initial vowel comes as failure imposes hopelessness...."

She emphasizes, however, that if you're going to understand what your cat is saying, you've got to know your cat: "The voice of one cat can be distinguished from that of another chiefly by means of the difference in their normal initial vowel-sounds, and until this norm has been determined for a cat, attempts to ascribe significance to particular shades of its vowels have little meaning."

Once you get the norms straight, then amazing insights are possible. In her second principal class of vocalization, Vowel Patterns—"sounds produced while the mouth is opened and then gradually closed"—there are the whispered or silent demand, the Begging Demand ("when in pursuing some end the cat becomes so completely absorbed by the goal that all else is lost sight of"), Bewilderment ("the *'maou?*... of bewildered failure of understanding" and "the tenser *'mæou?*... of worry"), the Milder Forms of the Mating Cry, and the Anger Wail. As the attentive cat person surely knows, the category of Complaint allows for quite a bit of variation. Cats' *complaints,* when judged to be insufficiently answered, can turn into *demands.* Moelk finds that she cannot always distinguish between certain "vaguely formed complaints" and "the more hopeless forms of demand," but she offers the comforting observation that "When complaint forms alone are used, the cat can very often be satisfied with mere verbal sympathy, which is not true of the demand." That must have been an awfully nice cat, to have been satisfied with verbal sympathy.

The third primary category of cat talk is what Moelk terms "Strained Intensity Patterns"—the mouth open and tense, the voice strong. Among them are the Growl, the Snarl, the Mating Cry ("a much modified form of the Demand," with many variations), the shriek of sudden pain, the Refusal (the "low, raspingly discontinuous sound sometimes uttered as the cat draws back in refusal from something urged upon it"), and Spitting, *fft!*

Kittens are born with the ability to make two sounds, a sort of quiet murmur, and the basic vowel pattern, shrill and insistent, sometimes beginning with an *m* sound, sometimes a *w*, which gradually becomes the familiar *mew* and then *meow*. They begin to purr at the age of two days. The "Level *'mhrn* of Greeting" comes into play during the third week, but its varieties denoting confident demand, request, and bewilderment develop much later—after ten weeks. Their mothers have their own language specific to motherhood. Most of it is spoken directly to the kittens. Whenever she's been away from them, she returns with soft little iterations of *'mhrn*—meaning, come to Mom to be licked. She will add a more commanding tone to the same sound when a kitten wanders farther away than she likes. She has a distinctive vocal invitation to play with her, a sound of frustration or irritation if the kitten stops playing too soon, a growl or even an anger-wail if the kitten breaks the rules—biting or clawing, for example. Kittens start to tussle with one another in their fifth week, and will spit in the midst of those bouts, but there's still no serious fighting, therefore no anger-wail or snarl.

At the same time she is in constant communication with her kittens, the mother cat also tends to talk more to her people. For one thing, she needs more food more frequently. Sometimes, also, she will seem to suffer from a certain postpartum loneliness or anxiety, and accordingly will solicit petting or expressions of sympathy and comfort from the humans watching over her and her new family. The more verbal interaction they have with people, and the earlier they have it, the sooner they will develop the finer points of communication, because only people have the awareness to encourage, narrow, and define particular outcomes to particular vocal expressions.

Cat talk is always interactive, and often aimed at achieving a goal. An individual cat's characteristic vocalizations can vary in a number of ways—among them loudness, persistence, and emotional intensity. The biggest influence on variation is probably the value the cat sets on the goal he has in mind—food, say, always a favorite. Getting you to open a door is another big one. Then there may be a matter of physical need, or instinctive pressure: *If I don't get to my sandbox this minute I'm going to burst,* or *I really reallly need to get out there and kill that pigeon.* Sometimes the cat may just be bullshitting you: *I haven't had my dinner, I'm starving.* Moelk points out that "time and repeated experience also build up varying degrees of anticipation with respect to different goals and different agents, and these various degrees of expectation are reflected in vocalization." Augusta, once she had achieved her ladyhood, would sit primly by her supper bowl and not say a word until you looked, and then she would offer only the softest, gentlest *mew?* Cats kept waiting, once they believe they've made their point but have not received a satisfactory response, will begin to change their vocalization, first laying a stronger emphasis on the first vowel of a vowel pattern, then increasing its intensity until the *request* becomes a decided *complaint.*

A cat's mood will certainly affect what she has to say, and the mood could well be left over from an unrelated experience: If she has just been pulled down from a tree that a dog chased her up, kitty will not speak in her normal tone of voice. An alteration of her typical sound could also be telling you she's sick. As a general rule, tone of voice is also an emotional indicator. The friendlier sounds are mostly higher-pitched, unpleasant sounds the low ones.

Finally, you may perceive something that you can choose to read as thanks, when, for example, you plop the old boy's dinner into his dish and a deep, satisfied purring ensues. Or you get, *This crap again? You know I hate it*—and a surly rasp of refusal.

It should perhaps be mentioned here that these "translations" are purely figurative: The internal experience of the cat is rather more basic. A cat's "words" are not symbols, abstractions, or names. They can vividly express the cat's feelings, however, and can certainly be specific in what they are requesting, enjoying, disliking, wanting, or needing. Elizabeth was convinced that Augusta asked for "milk" by name, and always answered her request with an echoing "milk!" We didn't know at the time that cow's milk isn't good for cats. Neither did Augusta, who, whether or not her little *mew* was a noun, certainly wanted that milk.

Not all cat talk is about want or need, however. Some is just nice. Augusta had her particular little gurgle-trill to say *Hello, hello, I'm so glad to see you and is this not a lovely world?* Goal-oriented? How often did that gurgle eventuate in a purr as she was talked to and stroked? She liked to be pushed flat on the floor and brushed hard, motoring like an Evinrude.

In one of Mildred Moelk's fascinating experiments, she set out to speak to every cat she met on the street for a month, "in a quiet, sympathetic tone," and stroke its head "when that approach was permitted." Out of the dozen cats she tried, only four declined to be petted. Not one gave her the *'mhrn* of greeting, which did not surprise her in the least, knowing, as she did, that cats reserve that word strictly for family and friends. She recorded the vocal responses of the congenial eight in International Phonetic Alphabet—mostly variations of *meow*—and kept careful notes of their behavior: the cat locked inside a store; the one who tried to lead her to open a door; the one sleeping in the sun ("near-purr behavior"); the one who answered in friendly meows, wanting to be petted but stuck behind a fence; the one sitting on a porch who responded only "after having been spoken to three times…who then returned at

71

once to staring toward the direction in which children, including two from his house, were walking away to school."

Even Mildred Moelk's detailed taxonomy of cat talk cannot describe the full range of what your own cat may have to say. That's because with the passage of time—especially when your cat has been with you since kittenhood—you and your cat together evolve your own particular way of speaking. Augusta, owing to her tendency to explore terra incognita, especially in Montana—where the supply of unknown places was essentially infinite—developed a number of calls to tell us she was lost and needed finding. The basic one was a loud *rowow!* much louder than one might have thought possible for such a little kitty. Augusta would guide our trudge through mudhole and down timber toward that sound with *aows* of gradually softening intensity as we drew nearer, and then, when she had located us and could be sure we were coming, the *aows* would quiet into pitiful infantile *miews*. (And there she'd be, up in a tree where we couldn't reach her, pacing back and forth on a high limb in a state of high anxiety. Eventually she would persuade herself to go back to the trunk, grab hold, swing down butt first hanging by her claws, and inch backward to safety—encouraged throughout by the you-can-do-its of her support staff.) She had a different, even more carrying holler for when she was lost in tall grass, for there she could not see out at all and could become entirely disoriented. The *rowow!* of come-get-me was displaced by an agonized *máa-oww*, falling at the end in resignation or despair.

As Susanne Schötz found in her acoustic analysis of cats looking through a window at birds and unable to get at them, a frustrated cat can produce a wondrous variety of sounds, trying trying *trying* to find the one chirr, chitter, chirrup, peep, pipe, twitter, warble, or quaver that will at last evoke the craved response. When Augusta sat at the screen door in Montana staring out into the merciless downpours of June, sometimes she would scold the rain, but as soon as she saw Elizabeth or me it became all our fault and she would try out hopeless variations of "please." "You're not a rain cat," we would say, because if we did let her out she wouldn't last a minute. Was she begging us to stop the rain?

Cats never address people in the savage vocabulary of courtship and mating. That is another language for another world. The female as she comes into estrus wails and moans like a victim of torture, and the competition for her favors induces the classic alley-cat chorus of groans, yells, snarls, hisses, and screeches. To the human ear it sounds as if the males are going to rip each other to shreds, and although they do sometimes fight, sometimes violently, the more typical dénouement of the hours of caterwauling will be the combatants' acceptance, however brief, of some one tom's first place in line. The loser typically will have signaled his defeat by crouching defensively, leaning away from his opponent, and lifting a forepaw. The female in heat will express her own acceptance of the winner's victory—as he mounts her—with a deep, muttered growl. (Her egg is not released until his penis enters her vagina.)

That is almost never the end of the scene, however. They may go at it again and again, as often as ten times an hour. Or number one may strut away in satisfaction, with the wailing, yowling argument of his fellows uninterrupted. Many female cats in heat will mate with a number of their supplicants. Hence the sometimes stunning diversity of kittens in a single litter: Each one may have had a different dad.[8]

℘

Cats express more with their faces than many people seem to realize. Certainly, compared to a dog's face, a cat's can seem uncommunicative. We have to learn to pay attention to subtle signs. It took me at least a couple of years of watching Augusta intently to realize that the interruption of a soft direct gaze by a slow blink is an expression of contentment. You should also know that staring straight into a cat's eyes can be taken as a sign of aggression or dominance, and may be met with unintended consequences. Only cats and their people when fully in each other's confidence can engage in the loving long gaze. Once you've reached that point, however, direct eye contact can develop its own subset of language. For one thing, you can now read the "eye whiskers" (not so easy with Augusta, black against black) and slight changes in the shape of the eyes. You may see longing, impatience, let's-play, anger, fear, need, love.

Some people say that when a cat comes into a room full of people whom he doesn't know, he inevitably seeks out the one who doesn't like cats. This may be true—again a matter of eye contact. The hypothesis is that the cat people all look at the cat, hoping for a response of some kind. The non-cat person will look away, wanting no part of it. The cat will then have taken all that staring as potentially threatening, and will perceive the person looking away as polite—just as unaggressive cats meeting for the first time will avoid eye contact. Hence the friendly approach to the wrong person.

Another friendly gesture, usually reserved for the cat's closest friends only, is the flick of the tongue from between closed lips, so quick sometimes you're not sure you've seen it. The position of the whiskers is also expressive. Curled forward and fanned out, they say that the cat is fully aware, unafraid, ready for action. When she is calm, they will stand straight out and come closer together. When she presses them back flat against her cheeks, watch out—that's anger or fear, and claws could be next.

A cat's mouth doesn't have a great range of expression—except, of course, when it's talking. A big yawn can be taken as an expression of comfort. Mouth agape, as we've seen, means flehmen—that special, hypersensitive mode of smelling other cats' pheromones. Exposed teeth with lips drawn back are a bad sign—sometimes dangerous to children who mistake that expression of fear and potentially immediate aggression for a smile.

The eyes are usually a component of a more complex expression. Seen alone, they can be ambiguous. Narrowed pupils may mean anger, or just that the ambient light is bright—the reactions are the same. Unusually wide pupils can mean fear, anger, or the acute attention of the hunter. When Augusta greeted us on our returning home, or just to say hello in the morning, she would stretch her legs forward in a deep bow, keeping her hind legs just loosely bent, and close her eyes for a couple of seconds. A cat purring in your lap will often close her eyes in contentment. When Augusta was being stroked or brushed, the milky nictitating membrane would come out from hiding to cover half of each eye—pure bliss.

A cat's ears can be eloquent. With more than twenty muscles controlling them, they can swivel an entire half-circle. Between cats, they are quick and easy indicators of mood. Lifted and forward means contented and maybe also ready to play—though in conflict situations the ears will take the same position, pointing directly toward the perceived threat. Straight up is relaxed attention. Ears out flat plus dilated eyes means ready to fight. Flat back with head down means scared. Swiveling, twitching, switching back and forth between positions all can indicate states of awareness, uncertainty, sensory overload. Sometimes a cat caught in ambiguity will have one ear wide open and the other closed flat. A cat will almost always turn her ears toward you when you speak, though she may otherwise seem not to notice. When both are open and cupped forward, she is maximally alert and aware. To see Augusta enter a woodland with her little black ears straight up, almost quivering, with her eyes wide open and every sense electric, was to touch the animal spirit in ourselves.

Posture and other body movements indicate a cat's fundamental mood. A frightened cat will draw himself up with an arched back, sideways to the potential aggressor, every hair on end, tail straight up—the classic Halloween silhouette—and sometimes crab-walk cautiously away, still sideways, saying, "I really don't want to fight." Deeper fright will turn the tail into an inverted U, sometimes curling forward between the legs. As tension mounts, the cat's breathing becomes more rapid, and the body becomes lower in back than in front. He seems to be making himself as small as possible. In utter terror, especially when cornered, the cat will take a defensive crouch, ears flattened, tailed curled tight around the body, eyes dilated to the maximum, whiskers back, every muscle quivering with tension, the only hope a last spring and full attack.

Anger—which can come on suddenly, even in the midst of a round of happy petting when the cat has had enough—is telegraphed first by a lashing tail. It will often seem incomprehensible to the person—*Where did this come from, what did I do?* "No one Cat is wholly good or bad," wrote T. S. Eliot in his poem "In Respect of Felines." "For even the nicest tabby that was ever born and weaned / Is capable of acting, on occasion, like a fiend."[9] A cat warns of aggressive action with stiff, slow steps, twitching ears, a relentless stare with pupils constricted to slits, the lashing tail. When he is about to attack, his ears will swivel back flat against his head, to avoid injury. Fortunately, real attacks on people are extremely rare.

A happy cat almost always approaches her people with tail held high, sometimes with the tip curled into a question mark. A cat minding her own business, content but feeling no need to communicate, usually carries her tail more or less straight out. A droop probably indicates a less than sunny mood. When a cat lying on her side grows irritated but isn't yet ready to take her leave, she will thump her tail hard on the floor. A cat sitting on her haunches with all four feet together and her tail wrapped softly around them is in a state of relaxed attention. Lying in sphinx position with paws tucked in and tail quietly alongside her body means peace of mind. A gently waving tail-tip usually indicates that the cat is listening intently but without tension. When the cat is sneaking up on prey, belly to the ground, the tail vibrates with excitement. Grown cats sometimes chase their own tails in a whirl of craziness—just for fun, it seems, though sometimes it may be in pursuit of a devilish itch. The tail is, of course, essential in the cat's extraordinary talent for balancing, as a counterweight that can change position in an instant.

Some cats will roll onto their backs in a number of circumstances—a gesture that some people confuse with a dog's belly-up posture of submission. Sometimes, not very often, it may signal a desire for a tummy rub. Other cats, apparently just as relaxed in the same position, will claw the bejesus out of you if you so much as touch that inviting soft fur. Females in estrus commonly express their readiness by rolling over. Some cats love to sleep on their backs, legs akimbo, evidently trusting that no harm will come to them despite the vulnerable position.

Augusta was never that confident: Her default sleeping posture was curled up tight, head sometimes buried in her side or behind her tail, not infrequently with one eye slightly open—just in case. When you spoke to her in that position, even when she seemed to be fast asleep, she would acknowledge your utterance with a soft curling of the tip of her tail. When brightly awake and full of beans, Augusta often did what we called "wiggle-worming," writhing at our feet, or "doing a banana," in which she stretched all four legs as far as they could reach while arching her back like a bow. Both were among the friendliest of her postures, with a modest, unassertive begging component mixed in. She certainly had it figured out that both gestures got our amused and affectionate attention.

Wiggle-worming in dirt or gravel seems to be a sort of self-grooming, but it also often denotes a readiness to play, especially if there's going to be chasing involved. Scratching something vertically leaves an olfactory Kilroy-was-here for other cats, and also aids in the shedding of old claws—with the new, very sharp ones just beneath. Trees and scratching-posts are ideally close at hand, but the furniture of many a cat owner can attest to their cat's persistence in marking his home territory no matter how many times hollered at or even bopped on the nose. A very gentle nose-bop emulates the *no* command of the mother cat. Anything more forceful than that is abuse. It won't work, anyway. Hitting a cat can only make him angry or afraid. Altogether too much trust is shattered when cat owners lose their tempers, and the distrust consequent to a single episode of loss of self-control can last life-long. Water pistol or spray bottle squirting can be effective, but they should be resorted to as seldom as possible, for an intelligent cat will sense the anger behind them. The result may be less an obedient cat than one who runs from the sight of you.

As in virtually every other aspect of their lives, what cats say to people shows a wide range of individual variation. Some cats are simply unexpressive—call it private, if you like. Some, notably some Siamese, can never shut up, and seem to be complaining much of the time. In between those extremes, what you say to your cat, what tone you say it in, how often you repeat it, how well you imitate the cat's own vocalization, and especially how early in the cat's life you start babbling at her all have an influence on her own loquacity. Like Bashan in Thomas Mann's "A Man and His Dog," cats like to hear their names again and again.* "Augusta, busta busta busta," we would say and say, "Auuu-*gusss*-taaah!" and as soon as

---

* "And what do I say to him? Mostly his own name, the two syllables which are of the utmost personal interest because they refer to himself...I rouse and stimulate his sense of his own ego by impressing upon him—varying my tone and emphasis—that he *is* Bashan and that Bashan is his name. By continuing this for a while I can actually produce in him a state of ecstasy, an intoxication with his own identity...." (p. 461).

she considered her current preoccupation finished—she did not like to be hurried—she would come trotting down the hall or bounding out of the tall grass, gurgling happily.

Even given the multitude of ways in which cats express themselves, a theory of the feline mind remains elusive. There's still imperfect agreement, after all, on the human mind. The simplest way to think about what cats are thinking—until some miraculous new machine is invented that can look inside their little skulls and read it—may be not to search for the "thoughts" processed in their subjective experience but to rely instead on what results they get. Michael J. Owren of Georgia State University and two colleagues propose that science "revise the definition of animal signaling by replacing the problematic notions of information and encoding with the broader, yet better bounded and testable idea of communication as influence."[10] That would naturally include the responses of their human companions. While we should always be cautious not to over-interpret, there is increasing evidence that people understand more of what animals are trying to say than science in the past has tended to admit.

Zana Bahlig-Pieren and Dennis C. Turner's experiment (cited in chapter 1) compared the interpretive prowess of people familiar with cats and those unfamiliar. They were surprised by how well the people did who didn't knows cats. "This result," they wrote, "may offer some support for [the] notion that humans have been selected for an ability to predict and interpret the behavior and emotional states of other non-human species. Empathy and anthropomorphism play an important role in this capacity, as it presumably also did in the original process of animal domestication."[11]

Some cat owners may not realize that their feline companions' expressive ability takes longer to develop to full proficiency than has been generally thought, and that it can continue to grow throughout their lives. Cats do not reach intellectual and social maturity until somewhere between the ages of two and four years.

That was certainly the case with Augusta. At the age of one she was still flighty, distractible, and impulsive. She was a superb athlete, able to jump from the floor to the top of the refrigerator, or even a higher shelf, with no appearance of effort. Horizontally she seemed to fly like Supercat, limbs outstretched, her trajectory nearly flat, across unbelievable distances. She loved to knock things down. By the next year, however, she was becoming, dare I say, contemplative. She might *watch* a butterfly before pouncing on it. She never became as vocal as many another cat, but as she grew older she also grew calmer. We inferred that her infancy must have been chaotic and scary. With peace and comfort and the passing of time, she began to sleep more soundly, play more gently, and speak more often.

Augusta adjusted well to her narrowed habitat in San Francisco, but with her return to Montana every summer the pure joyousness of her true self poured forth: Her fur was sleeker, her movements more buoyant, her voice less shy. She hated being in the car, and the scent of motels. Whether in her carrying case or not, she would employ her let-me-out and help-I'm-lost at high volume, and in a room strongly redolent of disinfectant she would moan and mew in misery all night—but as soon as we arrived somewhere congenial and had stayed long enough for her to settle her wracked nerves, Augusta sniffed every edge, rubbed her chin glands on every corner, and mapped every door, high perch, and window, wanting out, wanting in, and, if it was Montana, wanting out out *out* to my wilderness! In each new place she generated new routes and new routines, and literally made herself at home. She had traveled enough to have concluded that that most identifiable of scent stations, her litter box, would mark her home, wherever it should be, henceforth.

Whenever we left her, she grew anxious. One house we rented in Montana had a long driveway that led to a bridge over the West

Thinking? Talking?

Boulder River—five miles from the ranch where we used to live—and one night, returning home from a dinner party, we saw her eyeshine in the middle of the bridge. This was hundreds of yards from our cabin—where could she have thought she was going? Had she been trying to follow us? Did she have some scent, or sense, of her old home upriver, and was she headed blindly there? She was shivering, cold, and scared, crying weakly when we picked her up. There were also times when she wanted to be alone, and she had a fine talent for identifying hidey-holes where we could never find her, some of them right in the house, some well out into forest or brush. Sometimes—not wishing to end a nap, perhaps, and not hungry—she refused to emerge or make a peep no matter how long we called. Then, in her own good time, she would appear, bright-eyed and crooky-tailed, and ask for dinner.

And sometimes she would try so hard to tell us something and we just couldn't get it.

Chapter Four

The Wild Animal in Your House

The idea of taming wild creatures to serve human uses came along very recently in our species' history. The most important of our animal familiars—horses, pigs, cows, sheep, goats, chickens, and so on, the barnyard crowd—began to be confined and bred for human uses only about twelve thousand years ago. Dogs have been around a lot longer, possibly as long as forty thousand years. Cats, as we've seen, are newcomers, the first fossil evidence of domestication dating only to five-thousand-odd years ago.

Then there's the question of what "domestication" means in the case of cats. All the other domestic animals differ radically from their wild ancestors. The leading scholar of cat ancestry, Carlos A. Driscoll, and his colleagues write in a study of cat genetics that our house cats differ from their progenitors only in "behavior, tameness, and coat color diversity."[1] Any domestic cat can interbreed with a North African wildcat and produce viable kittens.

The full genome of the domestic cat was published only in 2014.[2] The research found that genes for "aesthetic qualities such as hair color, texture, and pattern" showed strong differences with wildcats, but most of those had entered the genome very recently:

There are now thirty to forty genetically distinct breeds, but two hundred years ago there were only five, and those had been created not so much by human intention as by historic geographical separation. Our house cat's ancestor, on the other hand, which is to say the forebear of the African wildcat, was shown by mitochondrial DNA analysis to be 131,000 years old. There was also evidence of continuous breeding back and forth between wild and domestic populations all the way through the history of the domestic cat—which the authors refer to, in fact, as *"only a semi-domesticated species."* The primary evolutionary force the geneticists found was "selection for docility, as a result of becoming accustomed to humans for food rewards." Sounds cold, in a way, but sounds like a cat.

In another article, Driscoll and colleagues do a tidy job of summing up the likely story of how the wildcat became the domestic cat and still stayed wild at heart:

> Considering that small cats do little obvious harm, people probably did not mind their company. They might have even encouraged the cats to stick around when they saw them dispatching mice and snakes. Cats may have held other appeal, too. Some experts speculate that wildcats just so happened to possess features that might have preadapted them to developing a relationship with people. In particular, these cats have "cute" features—large eyes, a snub face and a high, round forehead, among others—that are known to elicit nurturing from humans. In all likelihood, then, some people took kittens home simply because they found them adorable and tamed them, giving cats a first foothold at the human hearth.[3]

Given how little they've changed from their ancestors, it shouldn't be surprising that cats are the only domestic animal that can survive on its own in the wild nearly anywhere. Pigs can manage, too, in some habitats—many fewer, however, than the ecotypes in which domestic cats can make a living on their own—but within a few generations the domestic swine begin to revert to a sort of wild boar appearance and behavior, and they readily interbreed with wild boar. Cats stay cats, and can live in deserts, jungles, grasslands, forests, and beaches—as well as parks, golf courses, shopping malls, back yards, alleys, and even Roman ruins. Which can be a problem.

The rapid growth of feral cat populations in recent decades is responsible for a number of ecological troubles, as we will soon see, but the other tame beasts—cows, sheep, horses, donkeys, goats, ducks, chickens—have had far more powerful effects on our world. Their influence has been felt, and growing, for twelve thousand years. They have made animal husbandry and large-scale agriculture possible, which have made it possible in turn for humanity to conquer the planet and to change its very nature. In a rather gloomier paper, Driscoll writes, "The world's species are going extinct at a rate one hundred to one thousand times faster than the historic 'background' rate, primarily as a result of habitat loss, which is itself overwhelmingly driven by conversion of natural habitats to agriculture. However, to date, no domestic animal has gone extinct. The consequences for the planet (as well as for humanity and its domesticates) have been profound, and have included the complete transformation of almost every natural ecosystem on Earth."[4]

So don't blame kitty! All she's done is kill mice in the granary and knock off some birds.

The most peculiar thing is, What's she doing here at all? *Six hundred million* of her, worldwide.[5] Of those, a few are actually trained: These precious few—after long and patient effort on the part of their trainers (more patient, obviously, than Elizabeth and I ever were with Augusta and the Spider Ball)—will sit on command, fetch, jump through hoops, go for walks like a dog. The other day I saw a cat out for a walk on a leash, in the park. This was a long way from the nearest car. A small crowd had gathered, to marvel. Then I noticed that the cat was not actually walking, he was lying down. Then I noticed that the owner had sort of hidden the cat carrier a few feet away.

They do use their litter boxes, though! Augusta used hers from day one. Here's the thing, however. A lot of them won't. They stand in the box and do their business just outside it on the floor. Or they stand in it but they back up against the wall vibrating their tail and squirt straight back. Or they do it on the other cat's bed. Or the sofa in the family room. Or your pillow. When a dog does something like this, you can punish the dog and the dog will get it. If you rub a cat's nose in her error, or swat her with a newspaper, guaranteed, you will have made the problem worse, and the cat will never forget.

Many wild cat species have their own version of the litter box—the midden—where they pee and poop over and over in the same spot. They will maintain a small network of these middens to mark territorial boundaries or crossroads. Territory is of profound importance to cats, and those violations of litter box rules can nearly always be traced to some territorial misunderstanding (not on the cat's part).

Sometimes a cat is just nasty, or will turn nasty, and that can be dangerous. For the most part it happens only very rarely and for no apparent reason, and so the outburst will scare the daylights out of you. My cat Isabel recently woke in my lap and exploded into my face in a murderous fury. Those claws, when intended to injure, can really injure. Isabel raked four deep cuts in my forehead, two of them slicing down through my eyebrows. Luckily I was wearing my glasses. I'm still not sure what happened. I think she may have been having a nightmare and hadn't fully awoken.

The teeth evolved for killing. The canines, needle-sharp, can slip into your skin as easily as a hypodermic needle—a dirty one, too, teeming with bacteria. You might be operating some kind of chase toy and he might just suddenly overenthusiastically overshoot it, and before you have any idea what's going on he's "killing" your hand. If somebody played rough with your cat when he was a kitten, rough play might not yet have been civilized out of him. You've heard the term "scratch your eyes out"? Its original source was cat behavior, and it's not just a metaphor. When you and your cat are learning to play with each other, it pays to be careful till you're certain he knows the limits. Even then it still pays to be careful because once in a while, overexcited, he may forget those limits.

Besides inflicting physical hurt on your person there are also manifold ways in which cats can wreck your peace of mind. Ripping up the upholstery is a favorite. Eating plants. Knocking things off dressers. Deciding that the food that's been delicious for the last six months is now poison. Eating and eating and eating, with all the illnesses inevitably attendant. Cats may display all kinds of obsessive-compulsive behavior. Then there's yowling at four in the morning. Attacking your feet at five. Vomiting at six. How about running away for three days and then just strolling back in like, What's the matter with you? Or running away and not coming back.

You bring her a roommate (feline, that is) to alleviate her all too evident ennui, and both of them hate each other—and it doesn't get better. You take Augusta to Montana, which she's going to love, a drive of two and a half days which by now she understands perfectly well—she knows where she's going—and in the car she either protests loudly from inside her carrying case or else, if released, roams the interior in search of an exit (frequently seeking it behind the pedals) and then paces the motel room all night, moaning. You invest in a hideous and expensive seven-foot cat tower that combines scratching post with hidey-house and viewing platforms and dangling catnip-scented batting balls—and Fluffy gives it one sniff, after which the thing is forever invisible to her.

The Tufts University veterinary school identifies several possible triggers for the sudden onset of irrational anxiety (the cat's, at first, soon yours too): "thunder phobia or separation anxiety—the fear that owners who have left the house for a brief period will never return—the arrival of a new baby, the permanent departure of a family member, rearrangement of furniture, persistent and loud noise from outside…relocation of the litter box…flapping curtains…"[6] Rearrangement of furniture? *Flapping curtains?*

There is a whole industry dealing with this stuff. The queen of it is Mieshelle Nagelschneider, who claims the title of The Cat Whisperer. There are a lot of other self-appointed cat whisperers, but Nagelschneider is the one who has published a book—titled *The Cat Whisperer*—and goes on television. (The whole "whisperer" thing started with the original Horse Whisperer, Buck Brannaman, who accomplishes miracles with horses—admittedly a species more amenable to influence than cats.) The king of cat whisperers is Jackson Galaxy, who has his own TV show, *My Cat from Hell*, and three books. Mieshelle Nagelschneider is rather glamorous— short dresses and platform sandals. She was born and raised on a farm on the high desert of central Oregon. Pierced, tattooed, and earringed to the max, Jackson Galaxy—born Richard Kirschner, on the Upper West Side of Manhattan—looks like a Hell's Angel from outer space.

One reason they have both been so successful in dealing with the behavior of problem cats is their recognition of the nature of the species. "House cats," writes Nagelschneider, "are not fully domesticated (and some seem to retain more of their wildcat ancestors' instincts than others). For behavioral purposes, it's better to think of them as half wild."[7] Galaxy, on what he calls the Raw Cat: "It's fair to say cats remain comparatively undomesticated altogether."[8]

Nagelschneider sets forth a three-part framework for her approach to transgressive cats, cleverly labeled C-A-T. C is for *ceasing* the bad behavior, which she accomplishes either with an "act of God"—a noise or other disturbance that the cat can't identify as coming from a person—or with a distraction that's necessarily enacted before the behavior: For example, if Beastie is clearly on his way to stealing a pork chop, you toss a ball across the room that you're pretty sure he'll chase. A is for *attracting* Beastie somewhere else—say, to a scratching post—and then rewarding him with a treat. T, which she insists is crucial, is for *transforming* his *territory*. That part can be challenging.

Turning a cat's territory into a genuinely congenial environment—especially if the cat lives strictly indoors—is something that Jackson Galaxy has put a lot of thought into. In his book *Catification* he has you turning your house or apartment into a cat's paradise, with platforms, runways, hammocks, cocoons, spiral staircases, ramps, shelves up under the ceiling, even whole dead trees. You can end up making your habitat look almost as weird as him. No doubt there are people who love this kind of thing, I just don't happen to know any. No question, though, cats think it's great.

As we learned from Eileen Karsh, having two or more cats is a piece of cake if you adopt them from the same litter. Adopting them as unrelated kittens at the same time, and preferably of about the same age, isn't perfect, but it's pretty good. Bringing stranger cats together in adulthood can be really problematic. Like all-hell-breaking-loose, cats-at-war problematic. With the arrival of social maturity between the ages of two and four, territoriality also arrives, as well as an instinct for social hierarchy. The difficulties that these give rise to have led a lot of people in the past to the erroneous observation that cats just aren't social. Their wild correlates, North African wildcats, are in fact not very social, and domestic cats who grow up without human company develop a relatively exclusionary kind of society: The adult males are solitary except when in pursuit of breeding females, while closely related females look out for one another and babysit one another's kittens; otherwise, they seem, superficially, to have little interest in one another. Yet even in their individual solitude feral cats will live together in considerable numbers—sometimes hundreds—in colonial populations, with elaborate systems of hierarchy and a delicate politeness that prevents nearly all conflict. Even the operatic wailing, hissing, and fighting of tomcats in sexual rivalry are more often rhetorical than actively threatening. Real injuries are inflicted sometimes, and deaths do occur, but both are relatively rare. In general, cats prefer peace to war, and respect good manners—which demand maintaining a certain cool distance.

But sociality with others of their own kind is complicated, and sometimes, as in the case of our own intraspecies sociality, they just don't get along, or barely do.

When I lived in New York City, I had a husky male Chartreux named Catfish who was emotionally so needy, or should I say so affectionate, that I couldn't keep him out of my lap. He was a great cat but he was driving me nuts, so my wife and I agreed that a close relative from the same breeder was probably the best choice for a new companion. Accordingly, we acquired Catfish's two-years-younger half brother Bubba. When he was still a kitten Bubba looked just like Catfish at the same age, but as he matured it became evident that he wasn't going to have his half brother's Muhammad Ali physique. From then on Catfish spent a good part of every day bullying Bubba—chasing him up and down stairs, biting the back of his neck and pressing him to the floor, stealing his food. Finally, one day, Catfish and Bubba got into a serious fight and we had to separate them physically. Next day, another fight, even worse—they were drawing blood. We kept them apart, in separate rooms, to cool off, but every time they were released they were instantly back at it. In the end the only thing that worked was to send Bubba into exile at my wife's office for several days, after which things were fairly peaceful, though the dominance-and-submission routine never really let up.

Nagelschneider has statistics to show how common the Catfish-Bubba situation is: "In single-cat homes, the chances of the cat being sent to a shelter for behavior problems are 28 percent. Add a second cat and it's about 70 percent."[9]

Kirsten Weir wrote a terrific piece for Salon.com[10] that illustrated well what magic Nagelschneider can perform without ever even seeing your cat. Weir's cat Thompson was very clingy and affectionate, but he also would bite the hell out of her—really really bite, to the point where she bought red sheets to hide the bloodstains. Nagelschneider's intervention comprised one hour-long phone call and four weeks of email followup. That was it.

The first step in Thompson's treatment consisted of Weir and her husband going cold turkey on all expressions of affection to the cat. No petting, no lap, nothing. Next came clicker training. Whenever Thompson did something good, Weir was supposed to click a little metal clicker instantaneously—so that Thompson couldn't fail to make the association—and give Thompson a crunchy kitty treat. This kind of thing doesn't work if there's so much as a two-second delay. Such is the instantaneity, or distractibility, of the feline mind. That's why the clicker instead of just the treat.

Then came a kill sequence with a toy, so that Thompson would learn where biting was appropriate. Weir used a bird toy—always a good choice—and whenever Thompson nailed it and "killed" it, she'd click him and treat him. There was to be no punishment for bad behavior, ever. The worst it could get would be Nagelschneider's favorite "aversive technique" of rattling coins in a can—an excellent way to stop any cat from doing something you don't want him to do.

Four weeks in, exactly as planned, Thompson was pretty much cured. Every so often, he'd get overexcited and Weir would recognize the predatory look in his eyes and the flattening of his ears, but a quick shake of the rattle would suffice to remind him that he was a good kitty now.

Nagelschneider makes her living with such individual cat-by-cat consultations, most of them conducted by phone and e-mail. It is impossible to determine what her success rate is, however. There's no question that her book shows a thorough understanding of cats, but there is also no way to evaluate how often her remote-access methods really work. All she publishes are testimonials of admiration—no statistics. On her website and in every other communication she issues, she refers to herself as "Harvard-trained," but she declines to specify when and in what fashion her Harvard training took place. (In fact, she declined to answer any questions for this book at all.)

Jackson Galaxy makes house calls, and in his book *Cat Daddy* he makes no attempt to conceal his history of failures and self-doubt. Nor is he shy about the price of his success. His television show, *My Cat from Hell*, almost always features totally out-of-control cats and Jackson on call to their troubled households, often getting bitten or slashed as he tries to approach the guilty party with soothing words and manner. Having introduced himself on every program as "a musician by night and a cat behaviorist by day," he arrives in a pink convertible to a heavy-metal soundtrack and always brings his guitar case, but he rarely opens it. If you Google "My Cat from Hell guitar case," one of the first things that come up is "Why does the 'My Cat from Hell' guy bring a guitar to the homes?" In fact, it carries toys, catnip, stain remover, and various other tools of the trade.

The camera usually finds the people looking dazed and confused. They're sometimes physically scarred, too. The cats can be shockingly vicious—Galaxy gets scratched or bitten a lot, sometimes rather badly—but despite the convertible and the tatts and the piercings and the almost overwhelming impression of total bullshit, he works real, true magic on these profoundly screwed-up creatures, with solutions derived from a deep understanding of cat nature. Sometimes the answers can be as simple as relocating litter boxes, providing an elevated platform where a cat can perch above people level, or giving space to a cat who feels crowded. Teaching people to use a toy that lets the cat go through an uninterrupted hunt-catch-kill sequence often works wonders for a cat whose hunting instinct has been bottled up for years. It can be glorious to see some poor cat's human companions absolutely light up when the cat stalks a silly little bunch of feathers on a string, pounces on it, bites it, "disembowels" it with raking back feet, and finally retires to a safe and comfortable place to groom, with eyes half-closed in calm contentment. They just never knew.

The troubles nearly always have their source in people's ignorance. Galaxy patiently shows them that when you pet a cat repetitively from head to tail while watching TV while paying no attention whatever to the cat's rapidly mounting overstimulation, and the cat's tail begins to twitch and then to lash and you still don't notice, well, if the cat jumps down and runs away you can count yourself lucky. More often, the poor little guy has been trapped in a life universally characterized by your inattention to his attempts to tell you something, and his tension level has been building and building to the point where he's wound up so tight that all he feels he can do is hide—and then when you reach under the bed to haul him out and ask him what, please, is the *matter* with him, guess what? He scratches you with claws fully out and a terrifying, terrified screech. And you say this has been going on for the last three years?

Galaxy will lie on the floor next to the bed and offer a tiny dish of food. Then he gives the long, slow blink. Then he gives it again. The cat has never seen a human speaking his language, especially saying this. Galaxy just stays there on the floor, quiet and still. Another blink. Finally the cat blinks back. They exchange slow blinks. Tick by tick, the cat's tight-wired muscles begin to unwind, and, finally he creeps to the bowl. Huge Jackson Galaxy still lies there, talking softly, cheek on the rug. The cat can't believe it. Galaxy brings in one of the owners and demonstrates the slow blink. The cat blinks back, and stays calm. Magic.

Galaxy shows you that instead of petting the cat the way you think the cat ought to be petted, you might try just offering a finger, not touching at all. Tentatively, the cat comes, and rubs his jawline along your finger. Over the next couple of weeks, the cat shows you the places under and behind his ears, the tip of his chin, the top of his head—you can scratch there, it's okay—his flank back to here but no farther, nope! You learn.

You learn to pay attention. This cat is far more responsive than you had ever imagined, to your gestures, to the tone of your voice, to your mood. Sometimes it seems he's reading your mind, which he's not, but still, when you see how he watches you, and all he seems to see, it's remarkable. You've begun to feel better yourself, even when you're not at home. The cat is teaching you—he's so relieved! He thought for so long you were completely stupid. Mean, too. By the time of Galaxy's third visit, the cat is sitting in your lap, purring.

Besides the TV show, Jackson Galaxy maintains a vast web presence. Part of jacksongalaxy.com is devoted to the Jackson Galaxy Foundation, which focuses on shelters—attracting volunteers, promoting better design, helping individual shelters raise money, training shelter staff in rehabilitation of injured and otherwise damaged animals. The site also offers links to Galaxy's Public Figure Facebook page, the Instagram site thecatdaddy, Jackson Galaxy Cat Mojo on Google Plus, and Jackson Galaxy YouTube videos. There's an endlessly rolling Jackson Galaxy Twitter feed. There are click-throughs to Galaxy's bio, press clippings, and book reviews, signups for the Jackson Galaxy email newsletter, past personal appearances, upcoming public appearances, and even a "Where Should Jackson Go Next?" section. There's a very good Q&A series on cat problems, some drawn from the TV show, some just funky little selfie home videos of Galaxy talking to you—e.g., "How to Keep Cats Off Counter Tops," "How to Make Cats Let You Sleep," "Does My Cat Have Separation Anxiety?"

And far from least, there is the Jackson Galaxy retail empire: Purple Paw t-shirts, TeamCatMojo bracelets, Jackson Galaxy Natural Bobbler, Tipsy Nip, Tickle Pickle, Jackson Galaxy hooded zip sweatshirt, Crazy Cat Lady travel mug, Rainbow of Paws flip-flops, *Autism Awareness puzzle piece shoelaces*…hundreds of these things. From the beauty of his work down to…flip-flops? It does seem a bit much.

Finally jacksongalaxy.com takes you to the outer edge, a place where the animal behaviorist's sincerity and confidence begin to mingle with New Age capitalism, and the oxygen grows thin.

Galaxy's sincerity is persuasive: "I'm not religious," he once told an interviewer, "but I firmly have roots in the spirit world, and I'm consistently learning from cats what lies beyond the tangible…. You've probably seen your cat stare at a spot on the wall, just stare at it. If you're totally, overly human about it, you're going to be like: Dumb cat. Right? What I think is that they are locked into an energy pattern in that corner of your room. And they are observing and taking it in. If you allow yourself to see that, it makes perfect sense."

Then comes the capitalism, in the form of Jackson Galaxy's Spirit Essences, $23.95 per bottle, two fluid ounces each. You can put them in your cat's food, rub them on her fur, or—"Misting the house or apartment is a great way to treat the whole household."

To pick just one Spirit Essence: The BULLY REMEDY "reminds the 'big cheese' that things are fine without needing to patrol everyone else, and that it is not necessary to dominate in order to live harmoniously." Also for sale are FERAL FLOWER FORMULA, GROUCH REMEDY, HAPPY TUMMY, OBSESSION REMEDY, SELF-ESTEEM REMEDY, SEPARATION ANXIETY REMEDY, and many, many more. And oy, the ingredients! Among the dozens: poison oak, mule deer, swallow-wort, campfire, mosquito, and, I kid you not, wind.

Yet—try to follow this—*Please note: No actual plant, animal, or gem matter is used in our formulas, only the energetic blueprint.*

The which? Oh, never mind. The complete pharmacopoeia can be found at www.spiritessences.com/category-s/1876.htm.

And yet, and yet: Jackson Galaxy knows cats in ways and with an accuracy beyond what I've found anybody else to possess. He has recognized—this seems so simple, but it's deep—that the source of a great deal of "crazy" behavior—the sudden outbursts of scratching or biting or panic that seem to come out of nowhere— is often nothing more than *overstimulation,* which, past a certain threshold, induces panic. It could begin with what is actually a good experience—slow, gentle petting, say—but if it's too monotonous, or it starts as tolerable in a sensitive place but it goes on too long, then, in an instant, *yeaow!* comes the explosion. It can happen when you're just playing with a toy. Too many episodes of chase, pounce, and kill may amp up a cat's energy to the point where more than the toy starts to look like prey—your hand may all at once have turned magically into a bird. It's not necessarily easy to learn to recognize overstimulation before it crosses the red line, but if you have a dangerously "unpredictable" cat, it can be worth the effort to keep a careful watch. Like: While you're doing something with the cat that really needs your attention, don't also have the TV on, or be texting your friends.

Galaxy is able to recognize in each individual cat its particular tolerance for particular stresses in part because he knows the whole repertoire of usual stresses. The principal ones are actually few. Both Jackson Galaxy and Mieshelle Nagelschneider know that you can hardly overemphasize territoriality as a determinant of cats' behavior. And since we can't smell the scent markers of their territorial boundaries, and we don't have anything like the same instincts in the first place, it's all but impossible for us to read the tension that begins to build in a cat whose territory is being invaded, distorted, or threatened. But careful observation over time, especially when you have more than one cat, can give you a sense of where—geographically, even pheromonally—each cat feels comfortable and where the zones begin to blur toward discomfort. Sometimes you can't just draw this on a diagram. Territory can be three-dimensional. Galaxy recognizes some cats as "tree dwellers" and others as "bush dwellers"— ones that like to be up on a shelf or the fridge, others that find refuge in a blanketed cubby. Providing appropriate habitat for both can make conflict vanish.

Sometimes what Jackson Galaxy accomplishes borders on the mystical. When he was younger, working in a shelter in Boulder, Colorado, and just learning the slow-blink greeting that works so beautifully to express calm and peace to a cat, he found himself in a fourteen-by-fourteen-foot room containing forty-five cats in stainless steel cages, all screaming. It was two in the morning, the cats had just been transferred to an entirely new and strange place, and a Rocky Mountain thunderstorm was under way. He set himself the challenge of calming them one at a time with what he had begun to call "The Cat I Love You...Eyes, opened but lazy: 'I'—slowly closed, 'Love'—and open again, 'You.'" The first cat wouldn't buy it. He kept trying. At last the cat replied with the slow blink, and Galaxy moved on to the next cat. And so it went, all night, from one to the next, until the sun was coming up and the room was silent. Forty-five cats calm.[11]

The question naturally arises, Could you or I do this? The extent to which mysterious communication with animals has been most vividly on display in my experience is in the horsemanship "clinics" conducted by Buck Brannaman, the aforementioned original Horse Whisperer. (The term "horse whisperer" seems to have been in occasional use in equine circles for some time—some say it first denoted a sort of Roma magician with sorcerous powers over horses,[12] some say the original was a nineteenth-century Irish trainer named Daniel Sullivan[13]—but the locution was made famous as the title of a novel by Nicholas Evans published in 1995. Evans himself has written, "Others have falsely claimed to be the inspiration for Tom Booker in *The Horse Whisperer*. The one who truly inspired me was Buck Brannaman."[14])

I have seen Buck Brannaman at the center of a tight, small round pen whipping a handkerchief behind a terrified, out-of-control horse—never touching the horse, but psychologically pushing him up to and past his limit of panic—the horse racing blindly around and around, faster and faster, frantic, sweating and foaming, slamming against the sides as though to break his neck, until at last he softens, slows, stops, and turns toward Buck, head hung, quiet. Buck lowers the cloth, ever so subtly softens his posture, looks down at the ground, and the horse comes walking forward to have his muzzle softly stroked. By the middle of the day, the horse, who has never had a human being on his back, has blankets swung over his head, has been saddled and mounted from both sides, and is trotting, loping, walking, backing, backing in a circle with Buck aboard, all without the slightest tremor of anxiety.

Horses like this, with a colthood and youth of unfathomable

fears and phobias—and probably cruelty—even once well broke, tend to be forever afraid to enter a dark horse trailer. Yet I saw this happen: At the end of the long afternoon, Buck sat in the driver's seat of his pickup and with hand gestures alone directed that horse to walk calmly up the ramp and into the trailer, and then just stand there waiting for someone to come and close the chain behind him. I would never have believed it if I hadn't seen him do it—and the same thing again, and yet again. You can see the Horse Whisperer in action in the acclaimed documentary film *Buck*.

Buck Brannaman is a patient and generous teacher, willing to go over and over the often imperceptibly fleeting, almost ephemeral, gestures and responses that go into his miracles. My sense, however—and that, I think, of everybody else who participated in his clinics at our ranch in Montana—is that you can learn a hell of a lot from him, but in fifty years of study under him you can never quite do what he does. And how many of us could quiet a roomful of forty-five screaming cats?

Okay, we'll never be able to do forty-five. But from Jackson Galaxy we can learn to calm one scared cat. From Mieshelle Nagelschneider we can learn possible ways of reconciling two cats who have decided to hate each other. The core of these behaviorists' expertise is their understanding of the species' wild nature—where lies its relative inflexibility, and where its relative flexibility. Varying from cat to cat. Which is why we find both Nagelschneider and Galaxy over and over, and at length, coming back to issues of... the litter box. Every cat person has to get over the *ewww* factor and realize that to every cat, without exception, the litter box is a very big deal. It's at the heart of the species' fiercely surviving, indomitable devotion to territoriality; and we must never forget that much of their experience of it has to do not only with how it smells—to them, not to you—but also with the finest points of its placement within a cat's territory.

Galaxy tells the story of an especially eccentric and intractable cat named Benny, who hated all change. That's a very catlike characteristic; it was just extreme in Benny. When Galaxy moved his household and cats to a much bigger apartment, Benny took one look around and started peeing in three specific places—none of them being his litter box. Jackson went out and got three new litter boxes and put them precisely in the new places where Benny had been peeing. Bingo, that worked. But Galaxy wanted an apartment that was more than just a cat latrine, so he set out to move the boxes gradually closer together until they merged into one. He moved two of them a couple of feet toward the third, and suddenly Benny was peeing all over the place again. "The trick," he writes, "became finding his 'challenge line'...something that became part of my toolbox from that time on—finding the place where comfort changes to challenge. Think of a child dipping his toes in the pool versus jumping into the water." He found that he could move Benny's litter box half a foot a day, while the crazy cat slept. "Six inches was comfort; a foot, unacceptable challenge."[15]

If someone or something scares your cat while she's using her box, she may never use it in that place again. If you change the litter—please, none of the perfumed stuff!—that may be the end of that. If you don't keep it clean, you're asking for trouble (though Galaxy says cleaning it too often can also be interpreted as an unfriendly act). Put a hood on it, and you're probably going to scare the kitty. No liners, please, they hate 'em. There should always be a quick and easy avenue of escape, preferably more than one. In a multi-cat household, the rule is one litter box per cat, plus one more. And if you are even thinking about one of those auto-clean things, please stop thinking about it.

Beds, they can figure out, and as you surely know, they'll often have more than one, and find new ones. Perches likewise. All sorts of places in their territory will have a certain mutability, according to...who knows what?—often, probably, some olfactory business that you'll never understand. But the litter box is pretty much non-negotiable. It's up to you to get it right—depth, type of litter, placement, everything—or, well, pay the price.

But this does not presume to be a cat-care book. *The Original Cat Fancy Cat Bible* is probably as good as it gets. If you really want to get serious about arranging your cat's—or, in this case, more importantly, plural cats'—habitat, nothing can beat Galaxy's *Catification*. If you've got real cat problems, then Nagelschneider's *The Cat Whisperer* is the goods.

Unfortunately there are a lot of mediocre cat books out there. The best way to know a good one is to see how well it recognizes the essentially wild nature of the beast—how thoroughly its assertions are grounded in understanding of cats' territoriality, their predatory instincts, the importance of scent in their perception of their world, their rich vocabulary of vocalizations and body language, the complexity of their social and antisocial qualities, their sensitivity to overstimulation, the deep sources of their fears and avoidances. There have been, recently, some books that purport to be grounded in science but make no attempt to understand the subjective experience of cats. Without that, you will never have the slightest sense of who your cat really is.

Cat whisperers seem to be everywhere, and there's no governing authority, no medical board, no university degree, to tell you which ones are genuinely qualified and which are totally bogus. Even the good ones, like Mieshelle Nagelschneider, don't have any sort of diploma or other certification—except for the kind you buy (see "tricks of the trade," below)—and even she won't tell you her success-to-failure ratio. You have to go on faith, or at least recommendations. The field is chaos.

For example, at www.thepennyhoarder.com/cat-behaviorist you can learn this: "If you like cats and are willing to learn the tricks of the trade, **you can make up to $100 per hour or more as a cat behaviorist.**" Why, my goodness, "This job could be a satisfying career or *side hustle*. You don't need a *college degree* or a license," and, the website goes on, "there don't appear to be any laws regulating the cat therapy industry. But to boost your professional image, you *can* get certified by one of several organizations.

"For example, the International Association of Animal Behavior Consultants (IAABC) will certify you as an animal behavior consultant if you can submit evidence of sufficient experience and meet a few other basic criteria, including scoring 80 percent or higher on an application test. You'll pay a nonrefundable application fee of $125 and annual dues of $110.

"If you lack experience to get certified, you can at least join an appropriate organization for credibility. IAABC charges just a $50 application fee to join, and then $65 for annual dues. For this you are allowed to use the IAABC logo on your marketing materials, as long as you only present yourself as a 'member,' and not as a certified consultant.

"Study on your own, invest a little to be a member of an animal behaviorist organization, buy a few business cards, and you're ready to go."

Heed the language. "Tricks." "Hustle." "You don't need a college degree or a license." "Boost your professional image." Augh!

Yet a lot of cat owners can really use some help. Especially for households with multiple cats acquired at different times, especially when the house may not have adequate space for territorial comfort—or when the people just don't how to arrange what territorial space they might have—advice from a professional expert can be worth real money. If you've experienced any of the nightmare scenarios I've described (and there are others), you know how bad it can get. These unmanageable cat situations can destroy a whole family's peace of mind. The cats themselves, it hardly needs saying, are wretched as well. Fortunately, more and more often, a good veterinarian or your local SPCA can refer you to a legitimate professional. And even without books of their own or Spirit Essences, they can be remarkably successful.

Achieving real harmony with our semi-domesticated feline companions—something resembling actual domesticity—can be difficult, obviously. In the case of Augusta, we did not realize how difficult domesticity was for her (or, shall I say, how poorly we had reconciled her to it) until we recognized how happy she was in the woods and deep grass of Montana, where she could occupy her natural selfhood fully. In the gloss of her coat, in the arc of her bounds, in the brightness of her yellow-green eyes, in the fluid grace of her body in entire integration, you could not but see her delight. Certainly she could not have been here had she not been rescued from the nether side of the summer she was now taking such pleasure in, had she not been fed and loved, stroked, brushed, sung to, driven here in an automobile. With time, and care, and closer attention, she became a happy kitty in the city, too, racing upstairs, chasing a ball, attacking her rattly Anchovy Mouse, batting down the Furry Spider on its elastic string, squeezing through the door and grabbing ribbons through the windows of the Hotel Augusta that I built for her from a big cardboard box. Bathing, with the tiny white star on her belly showing. Snoozing on the loveseat while we watched a video, one ear flickering now and then, assuring herself that we were there—her family.

Imperfect, as families must be. But she was teaching us, showing us her nature. She taught me that if you want to have a wild animal in your house, you can have…a cat.

There is a crazy strain in American life, however, another of our inexplicable subcultures, that thinks you can bring the true, pure, undomesticated wild into your house and tame it. They think it's okay to keep wild animals as pets. There are thousands of these people. I used to go to a copy shop in Livingston, Montana, where from behind the counter an enormous gray wolf would rise, with a cocked head and a friendly look. He was the real thing, *Canis lupus*, not a hybrid. The owner had to tell her customers over and over not to try to pet him. He probably wouldn't have bitten anybody, really, but if he ever did, it wouldn't have been pretty. Families will raise a wolf pup with their little kids and will tell you, truthfully, that he has never harmed a hair of their heads, nor anybody else's. But then one day a visiting kid will smack one of the family's kids, hard, and the wolf will kill his packmate's assailant. Just defending the pack, y'know, doing his family duty.

It's understandable that people seek a certain uniqueness in a pet. In the last two centuries there has been a lot of breeding for certain appearances in domestic cats, and some of that breeding has included the introduction of wild genes, most notably for a breed known as the Bengal, which was accepted as a pure breed by the guardians of American cat purity, the Cat Fancy Association, only in 2016. Elizabeth's and my cat Isabel is a rescue mutt, but she alone in her highly various litter showed many characteristics of the Bengal, a breed descended from a cross of the Asian leopard cat, *Prionailurus bengalensis,* with the domestic cat, and bred through five or six successive generations toward ever greater domestic characteristics. (The original purpose was research into feline leukemia, to which the Asian leopard cat was immune. Though the hybrids didn't inherit the immunity, they turned out, almost miraculously, to be pliable, friendly, and tame.) Mongrel Isabel may be, but she's taller in back as Bengals are. Her voice is more often the Bengal *ack-ack* than the pussycat *meow.* She's spotted rather than striped, and her silky, flattish pelt shimmers in sunlight as though she has been sprinkled with gold dust—a very

Bengal look. She also, however, has the white feet and white belly of an old-fashioned brown tabby. Oddly, she has tufted ears (a mystery—Asian leopard cats don't have them), which among high-class Bengal breeders are considered officially Undesirable.[16] The quest for uniqueness, in strange ways, becomes a quest for conformity.

The forty-one pedigreed breeds "recognized" by the Cat Fancy Association, some much more exotic-looking than Isabel, have resulted from intense selection strictly within the domestic cat species, *Felis silvestris catus.* Of those forty-one, sixteen are recognized as "natural" or "foundational," and they tend to have fairly strong genetic diversity—because they are derived from random-bred forebears[17]—which gives them resistance to disease and makes their offspring unlikely to be deformed or otherwise unfit. Just as in the sad case of so many drastically overbred and inbred dogs, however, recently developed breed standards for some purebred cats allow what is known as line breeding, in which direct relatives, such as father and daughter or mother and son, are mated— with the result that any genetic defects spread rapidly throughout the breed.[18] "In any endeavor involving genetics," say the C.F.A. guidelines primly, "it is advisable to be very careful."[19]

In fact, breeding for uniqueness of the kind that's rewarded in cat shows—and implicitly encouraged by the Cat Fancy Association—has produced a host of genetic nightmares. Cats bred to be blue-eyed and pure white are likely also to be stone deaf. The genetic modification that creates the floppy ears of the Scottish Fold "also causes severe abnormalities of the cartilage of bones— there will be defective bone development and severe bone and cartilage abnormalities in all cats with folded ears. These bone abnormalities lead to severe and painful arthritis."[20] Maine Coons and Ragdolls are exceptionally vulnerable to hypertrophic cardiomyopathy, an ultimately fatal thickening of one of the two chambers of the heart.[21] Abyssinians and Somalis can inherit disorders that cause them to go blind between the ages of twelve and sixteen months.[22] Siamese can inherit a host of problems: crossed eyes, asthma, lymphoma, intestinal adenocarcinoma, chronic projectile vomiting, obsessive-compulsive disorder.[23, 24] Good old non-fancy mutt cats, the kind that shelters overflow with, are likely to be a whole lot healthier.

The quest for uniqueness gets worse. There are not a few breeders in search of the outlandish and the downright weird. One Judy Sugden, of Covina, California, has for thirty years "spent most of her time trying to concoct a cat with the temperament of a lapdog and the appearance of a tiger."[25] This kind of freelance what-the-hell experimentation has engendered God only knows how many monstrosities, moribund, crippled, racked with pain, infertile, deformed—"twisty cats,"[26] "munchkins," "squittens" and "kangaroo cats" (the last two being bred to have useless front legs so that they must hop like squirrels or kangaroos[27]). Many experimental kittens are simply born dead—to their Doctors Frankenstein a disappointment, to the decent world a mercy.

Then there are the people who think it's a fine idea to breed *F. s. catus* with this or that great-looking wild species and just see what happens. The Chausie, a cross between the jungle cat of southern Asia, *Felis chaus,* and the domestic Abyssinian, can weigh in at thirty-five pounds. Some of them, after generations of breeding with domestic cats, will be reasonably tame, but having stubbornly maintained the short intestinal tract of their wild ancestors, they're also, when fed cat food instead of wild prey, prone to food allergies and inflammatory bowel disease.[28] The Savannah is a hybrid of the African serval and a house cat, and like the Chausie is big, up to thirty pounds. The breed is illegal in a number of states, as well as New York City, and has been banned altogether in Australia, as "an extreme risk to native animals and the environment."[29] Jag cats are a stew of three wild species: bobcat, jungle cat, and Asian leopard cat. "Although all Jag cats are considered large," says the breeder Mokave, "some become gigantic! However, it is impossible to determine which kittens will become the biggest because full maturity takes four years and even tiny kittens have grown up to be giants!"[30] One infers that this breed has not been around very long, and wonders what, healthwise, say, the harvest will be.

Perhaps even crazier are the people who keep pure wild cats as pets. One of the species thus honored is the caracal, which in olden days was, in fact, "kept" in India and Persia, and trained to hunt hares and birds,[31] but in modern life is a thirty- to fifty-pound keg of feline dynamite. "They are not a pet of which a human can initiate affection at any moment," warns a guide to their care. "When caracals do play, they are rambunctious and destructive with average household objects and furniture."[32] Ocelots, being extraordinarily beautiful, are tempting—Salvador Dalí had one, named Babou—but they also "will not pay attention to disciplinary commands and have a pungent odor."[33] Bobcats, which look so cute in the wild, are all but impossible to tame.[34] The fishing cat of Southeast Asia is another that comes up in discussions of possible wild pets. Should it matter that it's an endangered species?

An organization called Big Cat Rescue cares for abandoned and abused big cats—lions, tigers, mountain lions, and such, often veterans of circuses and ghastly roadside attractions. Recently, Big Cat Rescue has been "seeing an alarming escalation in the number of hybrid cats being abandoned by their owners," but they have been able to take in only a fraction "because the problem is too vast." Why are so many being abandoned? "Genetic defects…cannot properly digest their food…projectile diarrhea…they bite…spray…loud howling throughout the night."

And what happens when they're on the loose? Big Cat Rescue explains: "Hybrid cats are much better hunters, due to their recently wild genes, and thus can do much more damage to the eco-system than feral cats alone. Add to that the likelihood of breed-ing with the feral cat population and you end up with much larger cats, capable of killing bigger and a wider array of native wildlife— including amphibious species because wild cats will readily go in the water after prey. Introducing wild cat traits into the feral cat population also imbues them with the wild cats' enhanced ability to evade humans, avoid traps, cross rivers, and travel much farther distances, which can spread the devastation into pristine areas that do not currently have feral cat populations."[35] Yet another horror: uncontrolled hybridization of feral domestic cats and escaped wild cats.

The great wild-pet-cat disaster of all time came to pass on Octo-ber 18, 2011, in Zanesville, Ohio, when a man named Terry Thomp-son, newly released from the penitentiary and newly informed of his wife's adultery, opened all the gates of his pets' cages and then shot himself in the mouth with a Ruger .357 Magnum pistol. By the time the police and their helpers from the Columbus Zoo had completed the only possible solution to the problem that Thomp-son had unleashed, eighteen tigers, seventeen lions, eight bears, three mountain lions, two wolves, one baboon, and one macaque lay dead.[36]

How could it have been possible for one household to have amassed such a menagerie? Wasn't it illegal to own a lion, a tiger, a grizzly bear? In fact, no, at least not in Ohio. They weren't even very expensive. You can buy a young lion for three hundred dol-lars. Adult lions and tigers "are effectively worthless," wrote Chris Heath in an article about this catastrophe in *GQ* magazine, "because there are usually more people trying to unload them than wanting to purchase them."[37]

Among the professional volunteers helping track down the animals was the Columbus Zoo's director emeritus John Bushnell

"Jungle Jack" Hanna, who is well known as an animal expert, having appeared on a number of national television shows in his safari outfit, always accompanied by animals from the zoo. He had once maintained his own private collection of wild pets, until one of his lions tore off a three-year-old boy's arm.

Besides the danger, consider the cruelty. It's bad enough seeing lions and tigers in zoos. In the worst of them the poor cats pace up and down and up and down all day in a concrete box, long since out of their minds, and even in the better ones, where they live in artificial "habitats" meant to emulate their native homes, they're not fooled. When you make a pet of a caracal or a jungle cat—or a leopard! some people own leopards!—you extinguish the light of a soul. Go online and you can find pictures of these mighty cats purring on the chests of their masters, or curled up asleep baby-peaceful on a blankie, but you won't see pictures of their reeking wire cages or the curtains and upholstery they've ripped to ribbons or the twenty-two stitches up the proud owner's arm. A horse's or a dog's it might, but a camera cannot photograph the broken spirit of a cat.

The noted wildlife biologist John L. Weaver had an idea for how to observe an animal in the wild with greater precision and intimacy than had ever been done before. After considering a number of species, in June 1992 he purchased from a fur farm near Flathead Lake in Montana a nineteen-day-old Canada lynx kitten. His timing was carefully chosen: She opened her eyes for the first time the next day. "I wanted her to imprint on us (Weaver, his wife Terry, and their daughter Anna), to see us immediately as her family." At home on the outskirts of Missoula, on a suburban lane that backed up to the Rattlesnake Wilderness, they bottle-fed the kitten every two to three hours around the clock. She was almost never out of their hands, and she slept in their beds, but Chirp—so named for her friendly greeting vocalization—though gentle and never aggressive, was emphatically not becoming a house pet. By the time she was ten weeks old she was spraying every corner in sight and tearing the furniture to shreds.

Weaver built a kennel of chain-link fence panels in the back yard, thirty-two by sixteen feet horizontally and eight feet high. "I wanted to test how quickly she was developing her prey-catching ability," he remembers, "and so one day when she was still just ten weeks, I put a live-caught snowshoe hare in the kennel. Three times her size. She caught and killed it exactly the way you expect a lynx to do—a quick bite to the back of the neck."

A measure of the biologist's success in raising the lynx to accept his companionship was that he could open the kennel door and stride straight in without a by-your-leave, and—adolescent now— twenty-eight-pound Chirp would fly through the air and wrap herself around his shoulders like a limp lynx-fur stole, purring loudly. He would give her full-body massages that lasted fifteen to twenty minutes (and she would reciprocate by methodically licking his face and head all over). His objective was by no means to make a pet of her but rather to habituate her to his physical proximity to a sufficient extent that when she came to adulthood he could mount a radio collar around her neck and follow her in the wild as she hunted. It would, he knew, be a punishing thrash through dense spruce-fir forest, with low needled limbs and brush raking his face as he plunged forward through the snow to keep up with her—holding a three-foot antenna aloft at the same time—but in keeping her under such close observation, he would have by far the most detailed data ever collected on this rare and notoriously elusive species.

In her second winter, he took Chirp to the mountains, and she was everything he had hoped for, tolerating his following but paying no attention to him, utterly focused on her hunt. "I had always wondered about those long tufts that lynxes have on their ears," says Weaver. "Various people had had various theories, but none of them sounded right to me. Then one day Chirp dug out a crater in the snow and snuggled down into it so that all you could see of her were the top of her head and those two long ears. To a hare moving fast nearby they might well have looked like just another hare." He measured her ears and a quantity of snowshoe hare ears and still more lynx ears at a fur farm, and they matched up perfectly. "She was camouflaged, lying in wait!"

Weaver and some of his colleagues were developing noninvasive genotyping techniques, using hair to identify the species, sex, age, and various other characteristics of animals drawn to scent stations. Working with Chirp, he formulated a sticky, stinky goop—"Still secret!" he insists—that could be smeared on patches of carpet and nailed to trees and would last for months without losing its potency. "When I finally hit the right mixture, Chirp just went crazy over it—rubbing and rubbing and rubbing that gland, you know, that they have along their mandibles"—hence leaving a nice DNA sample of hair to be sent in for analysis. When they were finished with their tracking or scent experiments for the day, generally Chirp would just lie down and wait for Weaver to sling her over his shoulder and carry her down the mountain to his pickup.

Chirp became so gentle that Weaver began to take her to community group meetings and schoolrooms. "I'd make sure the teacher had these first-graders all quiet and sitting on the floor at one end of the room, and then I'd let Chirp come in. She would walk right among the kids, checking them out, sniffing and looking, not making a sound. The kids were frozen. They had been told—in no uncertain terms—not to touch her. I don't think anything would have happened, but I played it safe. Sometimes I'd hold her in my lap in such a way that I knew I could control her, and then I'd let the kids pet her a little."

One night of Chirp's fifth year, someone slipped into the Weavers' back yard and, for reasons unknown, opened Chirp's kennel. She was never seen again.

What would John Weaver say to a person thinking of raising a Canada lynx as a pet?

"Don't do it. Don't do it. Chirp nearly destroyed our house. All in the name of science, you know, but—well, no, a lynx? A lynx cannot be a pet."

ℱ

Just thirty-odd miles south of Chirp's former home, a woman named Barbara Roe sells lynx kittens for $1,750 each. She advises her customers that "after nine weeks we recommend declawing all four paws."[38] This is beyond cruel, even to a house cat. The American Humane Society declares, "Declawing traditionally involves the amputation of the last bone of each toe. If performed on a human being, it would be like cutting off each finger at the last knuckle."[39] In many cases the pain continues long after surgery. The cats usually lose their sense of balance and must re-learn to walk on their semi-amputated feet. Robbed of their best natural weapon, they often become much more aggressive. Claws are for defense; teeth are for killing.

Keeping—imprisoning—a wild cat for a pet is an act, actually, of alienation from the wild. Seeking that extra touch of the wild in a hybrid is the same kind of self-defeating dream. Having robbed your captive of the fundamental qualities of its wildness—its freedom to be a predator, to make its own way in the world it evolved for—you will never in fact see the wildness for which you took on the whole futile enterprise in the first place.

But when you have a domestic cat, that only half-domesticated beast, you have the opportunity to witness the wild close up, because it's *mediated*, with such civil accommodation, by the pact that forms between the two of you. So it was for Augusta and me. As she grew into full adulthood, we were together almost all the time, and where some writers look out the window or pace up and down to wonder what comes next, I would contemplate Augusta. It came to feel less like mere observation than communion. I believe that in some way she was letting me in. That could never have been possible with a wild animal, in the house or in the wild.

Inside that witty furball purring on your lap is a wild beast anyway, constrained only by her advantages and the affection between you. In the next chapter we will see how very wild, once estranged from human love and care, kitty cat can be.

Chapter Five

The Wild Animal at Large

The last white plastic trash bag whumps down and the silence is sudden. Last bedroom lights, above, are winking out. The street is remarkably dark. River mist dims the Portico d'Ottavia, slicks the basalt cobbles black, muffles the puncture and pull of crows' beaks ripping into garbage. Shadows slide along the walls, becoming cats. From the ruins of the Theatre of Marcellus, the Temple of Apollo Sosiano, scrambling up the pitted stone walls from the brush along the Tiber quays, out from between the security sensors of the Great Synagogue's bosky garden they come, all converging on this one short restaurant block. The crows flutter and give way; the cats paw in. Ba'Ghetto Milky has plentiful leavings of fish, BellaCarne of meat. No fighting, only the occasional hiss or swat—there is plenty for all—Daruma Sushi Kosher, Fonzie the Burger's House Kosher, Zi Fenizia Kosher Eatery, Antico Ghetto Ebraico Sheva, Il Giardino Romano, La Dolceroma, Giggetto....

All across the sleeping city, cats are in motion, deploying from their haunts in the ruins and parks to alleys, rooftops, gardens, windowsills, doorsteps, the plinths of monuments. There is food everywhere, here or there a run-over rat, mouse, or pigeon, dropped hot dogs, neglected spills of refuse, but all those together could never sustain these thousands of cats. What does keep them thriving are the ton on ton of commercial cat food lovingly placed at official, city-registered feeding sites by Rome's ubiquitous *gattare*, cat ladies. There are cat gentlemen, too, *gattari*, but few.

Care is what the cats of Rome get, not just food. Rome has long taken pride in its identity as *La Città dei Gatti*, The City of Cats. In its streets, and in the ruins of the ancient imperial city, live uncounted cats, most of them feral. Others of Rome's free-roaming cats are abandoned pets, or just lost. Estimates of the population size vary wildly, up to three hundred thousand. One hundred thousand seems more plausible, but they have never been counted. Even the biologists who study them can do no more than hazard a guess.

In 1988 the Roman municipal government forbade the euthanasia of any cat or dog except those deemed to be incurably ill or fatally injured. In August 1991 the Parliament of Italy passed a parallel measure, stipulating, "Feral cats have the right to live free… and cannot be moved from their colony," and also that they must "be surgically neutered by the local Veterinary Public Services and reintroduced to their colony."[1] Despite the ongoing campaign of neutering, the feral cat population is probably growing.

The biologist Eugenia Natoli (pronounced with emphasis on the *o*), who has been studying the cats of Rome since the early 1980s, had already foreseen a population explosion that would result in massive starvation, early kitten deaths, widespread disease among the cats, and the prospect of zoonoses (illnesses that escape from animals to humans). Working with the World Health Organization and the United Nations Food and Agricultural Organization, Natoli proposed that because many of Rome's feral cats were injured or otherwise unhealthy, the capture necessary to neuter them could also afford an opportunity to administer veterinary care.[2] This was going to require mobilizing and coordinating a welter of government agencies and, most important, an army of volunteers, drawn mainly from the city's plenitude of *gattare*.

It was going to be the most comprehensive program ever attempted to control a feral cat population by the technique known as Trap, Neuter, and Return, or TNR. Because not a single large TNR program had ever been carried out comprehensively, any- where in the world—that is, because not enough of the population had been neutered, or because the population wasn't sufficiently isolated and, therefore, new fertile animals continued to enter it—the record had been one of failure after failure. The colonies kept growing. "Trap, Neuter, and Re-abandon," its detractors dubbed it.

Rome's program was heroic. Natoli and her colleagues studied the city's feral cats for ten years, from 1991 to 2000. They moni- tored 103 colonies, whose *gattare* captured almost eight thousand cats to be neutered (of whom only two died). Each neutered cat had a corner of an ear lopped off for easy subsequent identifica- tion. At the end of the ten years, fifty-five of the 103 colonies had grown smaller, twenty had shown no change, and twenty-eight had actually grown larger. What was totally unexpected was the tremendous amount of leakage, both in and out. Cats ran away, and cats showed up—the latter usually abandoned surreptitiously by their owners, often accompanied by kittens. The kittens were fairly easily adopted out, but the colonies were stuck with the adult newcomers. If there had been no in-migration, the program would have been a success. But this was the real world, and, as Natoli's study concluded, "TNR programs alone are not sufficient for managing urban feral-cat demography, and we suggest that they be matched with an effective educational campaign directed to citizens to reduce the high risk of owned-cat abandonment."[3] That remains the Achilles' heel of no-kill shelters: the never-ending influx of abandoned cats, most often kittens.

As dawn spreads the city's distinctive golden light across the tumbled columns, the still-standing temples and arches, the unmown grass and climbing vines, the last of the street-prowling cats saunter home full-bellied to groom themselves in slices of warm sun on the marble, eyelids heavy. In the immense rectangular pit known as the Largo di Torre Argentina, dug into the heart of the modern city in the early twentieth century, tall umbrella pines and rampant weedy wildflowers grow amid the remains of four temples from the time of the Roman Republic—the oldest of them four centuries older than those of the Imperial Forum—plus the Theatre of Pompey, where on March 15, 44 B.C., a group of senators led by Gaius Cassius Longinus and Marcus Junius Brutus (Cassius and Brutus, in Shakespeare's rendering of the event) stabbed the emperor Julius Caesar twenty-three times.[4] As morning takes hold, Romans by the hundreds hurry past without looking. The Via di Torre Argentina is a major bus stop, and the Corso Vittorio Emanuele II, along the excavation's north side and, in fact, built atop part of the temple complex, is one of the few remaining heavily trafficked streets in the historic center (much of the rest of central Rome is a blessedly traffic-free pedestrian zone). Along the southern boundary, tram line 8 rumbles along its rails toward Trastevere. Shops and bars crowd all four streetsides, and along the wide western promenade stands the legendary Teatro Argentina, where *The Barber of Seville* had its premiere. Tourists come to the railing and peer down, interested less in ancient history than in the current occupants. Cats lived here when Caesar was murdered, and cats live here now, the most famous cats in Rome.

In a corner of the site is a steel-barred door, beyond which a steep steel stair leads down to a small, neat garden. On the door is a sign:

VIETATO DARE CIBO AI GATTI

THIS IS AN IMPORTANT ARCHEOLOGICAL SITE

PLEASE BE ADVISED THAT IT IS FORBIDDEN

TO LEAVE CAT FOOD IN THIS AREA

These cats are fed and cared for by an authorized organization

All of our cats are sterilised and vaccinated

THE ABANDONMENT OF ALL ANIMALS IS ILLEGAL

For information you can find us in the corner at the bottom of the
iron stairs

Another sign informs you that minors may enter only if accompanied by adults and that whatever it is down there at the bottom of the iron stairs is open seven days a week from noon to six. It also says, "Adoptions 1 p.m. – 5 p.m."

The Largo Argentina was once, like all the other ruins in Rome, full of cats in chaos, without veterinary care, never neutered, but always fed by Rome's sweet-hearted cat ladies. This particular population began to attract attention when the film and stage star Anna Magnani, who appeared often at the Teatro Argentina just across the street, joined the ranks of the *gattare*. After Magnani's death, three local women struggled to keep the cats cared for, but abandonments were out of control. In the dank half-open temple rooms beneath the Corso Vittorio, without electricity or running water, the *gattare* created a primitive shelter. In 1995 an organization called the Anglo-Italian Society for the Protection of Animals discovered their plight and began to pitch in with money and materials. The volunteers began to solicit money from the tourists drawn to the archeological dig. Wealthier and worldlier donors took an interest and taught the Roman *gattare* about big-time fund-raising— dinners, raffles, galas. Soon not only were the volunteers able to provide full care for the Largo Argentina's cats, they began to share their wealth with other Roman cat colonies. They opened a clinic to which owners could bring their cats for sterilization and medical care. An ancient tunnel was adapted as a refuge for sick or injured cats and for cats recently abandoned to the colony, so that the newcomers could be introduced gradually to the longer-term residents. Two books set in the sanctuary were published, and they sold well. Tech-skilled friends built a website, www.romancats.com. An English volunteer produced a film that the Largo staff distributed on DVD. A Cat Pride march demanded municipal funding for the protection and care of Rome's feral cats.[5]

Being a *gattara* was becoming fashionable. Fewer than half of Italian women work outside the home,[6] and it is an affluent country with a strong social conscience, so there were plenty of well-off women with the time and the desire to take care of the cats. Most of the volunteers today are notably well-dressed and prosperous-looking.

It remains a struggle to keep the population under control, because people continue to sneak in at night and abandon their cats in the Largo, knowing full well how nicely they'll be looked after. It's not seldom that litters of kittens turn up. It's well known as an adoption center: In the year 2013, 143 Largo Argentina cats were adopted; in 2014 there were 165 adoptions; in 2015, 148.[7]

It's not all that hard for a cat who wants to leave on his own to do so. In May 2015 I was walking late at night in the dark, narrow Via dei Funari southeast of the Largo Argentina when I came on a fresh white poster on a seedy, graffiti-covered wall, with a photograph of a bright-eyed gray-and-white tabby cat and this copy (translated):

KARAMAZOV
is a cat in the care of the Cultural Association of the
Torre Argentina Cat Colony. He loved to wander around

near the entrance of the refuge. Unfortunately we haven't seen him for several days. If anyone could give us news of him, we would be very grateful. Otherwise we must think the worst.

If you have seen him, please call us at [three phone numbers].

Thank you for your help,
The volunteers of Torre Argentina.

What a shame, I said to myself. I walked on, and two minutes later, in Piazza Campitelli, there was a gray-and-white tabby drinking from a tall Baroque fountain. No question, it was Karamazov. It was so quiet you could hear the soft mizzle of water from the high uppermost bowl to the great middle one down to Karamazov's shimmering marble pool. (The fountain is the work of Giacomo della Porta, and spectacular.) I walked toward him with my phone camera at the ready, as stealthily as I could. I had come within ten feet of him when he looked up at me like, What do *you* want? This was clearly a cat accustomed to humans. I managed to snap two quick pictures before he dashed away, tail high and curved in horseshoe shape, the full let's-play position. I found him waiting for me under a Vespa three minutes farther on, in Piazza Mattei, with its superb tortoise fountain (also a late-sixteenth-century design of Giacomo della Porta, though supposedly the famous bronze turtles were added seventy years later by none other than Gian Lorenzo Bernini). Again Karamazov let me come close enough to click another photograph before he fled, every ounce of him calling out *Chase me, chase me!* but I had to get back to that poster. It was late, and I knew I might be waking someone, but I thought they might want me to, and I was right. Did he speak English? He did.

—I think I've found Karamazov.

—Oh my God!

—I got a couple photographs of him.

—You can text one to me?

—Sure.

I did so. My phone rang immediately after.

—It's Karamazov! Where is he?

—Five minutes ago, he was in Piazza Mattei.

—Oh my God! Two minutes away from the Largo! It's a good place, no traffic. We find him tomorrow.

The next afternoon, I went to the volunteers' office and met the guy I had talked to, Daniele Petrucci. They hadn't found Karamazov, and all the volunteers were distraught. (I imagined how I would be feeling if Augusta had disappeared into the streets of San Francisco, and some stranger had called in the middle of the night to say she was just down the block, in a safe place, and now I couldn't find her.) Karamazov was a longtime resident of Torre Argentina, and beloved. It took a week for Daniele and his friends to find that cat. Karamazov had traveled into the Imperial Forum, the heart of ancient Rome—still an active archeological site, a maze of magnificent ruins, as well as the home of hundreds of cats. He had let himself be caught with no trouble, ready to come home.

But a year later, Karamazov moved on again, and once again toward the Forum. This time he took up residence on the Campidoglio, with its piazza designed by Michelangelo and, for shelter and cover, just enough of a little pine-shaded park below, from which he had (if he cared) a commanding view of the Forum of Caesar, including the ruins of the once all-marble Public Lavatories, the most luxurious public bathrooms in history.[8] Policemen assumed the feeding of Karamazov, and he became popular with the many tourists there as well as Roman regulars and, *naturalmente*, the neighborhood cat ladies.

This was a cat with charisma. When I first spoke with Susan Wheeler, the American head of Friends of Roman Cats, and started to tell her the lost-Karamazov story, she interrupted me—"I know Karamazov." I went on, how he was found in the Forum, and she said, "That was Daniele?" She was surprised to hear that Karamazov had moved house—he had been at Largo Argentina for so long, and was such a favorite there. To give you an idea of the intensity of attachment these cats inspire, Susan Wheeler conducts *cat tours* of Italy, visiting feral colonies in Rome, Florence, Venice, Arezzo, Montecatini Terme.

Eugenia Natoli observed that between the 1970s and 1990s, the *gattare* of Rome had greatly improved. They used to leave plastic bags behind, they didn't clean up uneaten food, they paid scarce attention to neutering and other medical needs—pretty much, they just fed cats. And when the population got too big, cats were killed. By the '70s the *gattare* were officially sanctioned, and registered with the government, so it seems unlikely that they themselves would have done the killing, but nobody knew who did, or nobody was saying.[9] Municipal oversight, however, and probably also the evolution of conscience everywhere regarding animals, brought real change. The city of Rome now publishes and distributes a "Ten Commandments of the Perfect Cat Lady (and the Perfect Cat Gentleman)." Paraphrased in English and shortened, they are:

1. Distribute food at fixed hours.
2. Create feeding stations in the shade to avoid decomposition and odors.
3. Serve food in disposable containers and remove them promptly.
4. Choose food rationally and not "whatever."
5. Assure that there is always water available.

6. Choose a safe place for shelter boxes and clean them often.
7. Avoid leaving food under parked cars.
8. Obtain the agreement of condominium owners for where food may be left for cats in courtyards and gardens.
9. Remember that these rules are insufficient if you have not seen to the sterilization of cats by the municipal veterinary agency.
10. Every colony is entitled to care under National Law 281/91 and Regional Law 34/97. Inability to protect cats constitutes the crime of mistreatment and is punishable by imprisonment from three months to one year or a fine of three thousand to fifteen thousand euros; the sentence to be increased by half if you cause the death of an animal.

In a study of the Largo Argentina colony and two other Roman colonies beginning in 1995, Natoli found that the volunteers were still bringing excessive amounts of food, but now they were taking away yesterday's leftovers when they brought today's feasts. They also cleaned up trash, and were so dedicated to humane population control that they were paying for veterinary care out of their own pockets even though by then Rome's Public Veterinary Services would provide it for free.[10]

So it was certainly not in quest of food that Karamazov had made his escape. Once on the loose, in fact, he may have been getting hungry. Urban feral cats, contrary to popular belief, are not great hunters[11]—there wasn't much to hunt, anyway—and the skinny local street cats and the Forum's established residents would have been guarding their resources fiercely against this sleek, well-fed intruder.

Was it that he couldn't stand the crowding? Natoli's study showed that in the 5,400 square meters of the Largo Argentina colony—about an acre and a third—the population density was equivalent to 14,444 cats per square kilometer, or just under 38,000 per square mile. That is a whole bunch of cats. But studies[12] have shown that where food is abundant, cats develop completely new systems of sociality that can allow for high density. Picture how New Yorkers and Parisians preserve their dignity on a crowded subway car.

In *Rome and a Villa,* Eleanor Clark wrote of watching the cats of Piazza Vittorio Emanuele—a very dense population indeed—hurrying to cover from an imminent cloudburst:

> The lines they trace are…all in a criss-cross as of night projectiles over a battlefield. It is each not to the nearest cover but the pre-determined one, which may be halfway round the ruin or all the way across the lawn, one way to some nook in the structure, another to the sheltering brown pantalettes of a pygmy palm, so for a few seconds the whole ground is a contradiction of flying cat-furs, which resolves itself in a moment without collision or argument or even a swerve of line unless a particular place were taken, when the late-comer shoots off by the same mystic geometry to the nearest alternative. Then there is no further move.[13]

Population density, then, surely, was not what prompted Karamazov to wander. As is so often the case with cats, his escapade offered little room even for a guess. He just went. When he was brought back, he showed every sign of being glad to be back. Then he left again, for pastures new.

People liked Karamazov because he seemed to like people. A fair number of the cats who end up in Rome's more privileged shelters have had some amount of domesticity and human companionship in their backgrounds, and will let themselves be petted and even picked up. Others are hard-core feral and look it, with gouged-out eyes or missing legs or patches of raw skin—unpleasant, scary-looking critters, whom one best not try to touch. Yet even they, once adjusted to colony life, almost all become peaceable and calm. Rome and Romans have a civilizing effect. You see it everywhere.

Take for example two Roman wildlife biologists. American wildlife scientists tend toward attire somewhere between safari and thrift shop, and usually need better haircuts. Eugenia Natoli dresses with elegant flair, tailored jackets, slim skirts, silk scarves, fine jewelry, high heels, just-so coiffure. Luigi Boitani, one of the world's most renowned wildlife scientists, is given to silky tweed, chic dark shirts, cashmere sweaters over the shoulders, a big scarf wound high up to the chin, soft Italian leather shoes. Perhaps Karamazov wanted a better-looking place to live.

I take a long bus ride to another refuge. The Pyramid of Cestius is a striking bright, white, steep one, sharp-edged and sharp-pointed, which rises from a bright green sunken meadow strewn with fragments of monumental marble, a good twenty feet below street level. If you wonder why it is that all the ruins in Rome seem to be so far down, the science writer David Quammen offers up Charles Darwin's idea:

> At least some…species of earthworm had the habit of depositing their castings above ground. A worm would back tail-first out of its burrow and unload a neat castellated pile around the entrance. As a result, Darwin recognized, soil from a foot or more underground was steadily being carried up to the surface.…Earthworms therefore were not only creating the planet's thin layer of fertile soil; they were also constantly turning it inside out. They were burying old Roman ruins.…[14]

Traffic whizzes by and the earth shakes as the Metro hurtles into the Piramide station, but the cats snooze in the shade and in the sun, oblivious.

Adjacent to the Pyramid is their nighttime prowling-ground, the leafy, cool Protestant Cemetery, where John Keats and Percy Bysshe Shelley are buried, not to mention the Beat poet Gregory Corso, the Italian Communist Antonio Gramsci, and Giorgio Bulgari, the diamond meister. Many of the monuments are fancifully ornate examples of the gravestone-maker's art, and the whole place is impeccably kept. Somebody, doubtless from among the *gattare,* evidently picks up the cat poop, because there's none to be seen. There must be a lot of it, too, for the Pyramid cats are very well-fed, some of them tubby even. Someone—the city of Rome? the volunteers?—has hidden audio somewhere in the ruins, so that at night the cats have classical music playing for them.[15]

In the dreary industrial reaches of Trastevere, Eugenia Natoli and I walk down a long, lonely, graffiti-walled street, to the *Colonia Felina Comunale di Porta Portese.* The colony is home, today, to 231 cats in what for feline Romans surely is idyllic luxury: a long blue pool filled with clear water constantly refreshed, blanketed beds, abundant and clean litter boxes, little private houses for cats wishing solitude, big ones for the more social, spreading old plane trees for shade and to climb, high walls to perch on and survey the world—either their own realm, full of their own well-groomed, civil friends, or the larger world just outside, featuring the ratty, untameable cats of the Tiber banks (who sneak in just before dawn to feast on the residents' leftovers). The Porta Portese compound is chockablock with cats, but the mood is of tranquility. Matilde Talli, in charge of the colony, is a soft-spoken older woman with a quiet, reserved manner and a radiant love for her cats. I watch as she calls to an old, stone-blind cat some twenty-five feet away and he at once comes trotting briskly, navigating around a tree and not falling into the pond, both of which are on the straight line between him and her. A couple of other cats step politely out of his path. Signora Talli scratches the blind cat's ruff vigorously, explaining, "He knows his way around by now, and the others respect him."

Starting in 1911, this was a dog shelter. Lost and abandoned dogs had three days to be claimed; when not, they were killed. With the passage of the national anti-euthanasia law in 1991, it became a true hell for dogs, if you believe, as I do, that for a dog solitary confinement is worse than death. Each dog was locked in a cage and never let out unless adopted, its world a ceaseless cacophony of barking, whining, moaning despair. In 2003, renovated, the compound became a humane cat shelter, and in 2014 it was transformed to its present cheerful and immaculate state. There are conscientiously tended flower beds, a granite mosaic terrace, fresh paint, a sense of orderly peace.

Porta Portese is a good place for abandoned kittens. Adoptions are frequent. There is a wall of portrait photographs of adopted alumni, in chronological order since the renovation and reopening: "N.3 Liliana," a sleepyhead; "N.4 Perla," a big-eared brown tabby, looking startled; "N.5 Birba," a sad-eyed dark gray; "N.6 Volkswagen," a gold-muzzled kitten; "N.7 Landini," a classic golden-eyed solid black; and dozens more. In addition to three full-time paid staff, Porta Portese has forty-five volunteers. That comes pretty close to one human for every five cats.

This extent of voluntarism tells you how hard it would be to replicate the Roman experience in the United States. Where would we ever get so many and such loving volunteers?

Colonies like these—Torre Argentina, Piramide, Porta Portese—demonstrate pretty conclusively that feral cats can live peacefully in colonies at very high density as long as their lives are made comfortable. Istanbul famously sustains a huge population of feral cats, as adored by their city as the cats of Rome are by theirs. The Turkish cats are not cared for with the official rigor seen in Rome, but there are cat shelters everywhere, some built from packing boxes, and bowls of food and water bearing signs of welcome—"Cat restaurant, bon appétit"—and warning—"If you don't want to be desperate for a drink of water in the next life, don't touch these cups." Istanbul's love of cats goes back thousands of years. During construction of a subway tunnel on the shore of the Bosporus, a cat skeleton was found with a healed broken leg and dated to be 3,500 years old. A zoologist from Istanbul University concluded that the bone could not have healed in the particular way it did unless the fractured leg had been wrapped by a human.[16]

The widespread assertion, even by self-proclaimed experts, that domestic cats are strictly loners, like their desert ancestors, is so false in so many ways it's amazing that it persists. There are orderly populations of cats all over the world. Whenever cats need to live together in groups, they form their own sort of civilization, with rules, boundaries, social strictures, dominance hierarchies, gender-related customs, and systems of behavior for protecting the young and their mothers. When hunger sets in, just as in human societies, social order may begin to break down. When resources are abundant, however, social harmony—the feline version, anyway—can become well established. A study comparing a feral cat population in rural France and a very dense urban one in Lyon tells how decent urban life can be (and also hints at how an inherent, and flexible, talent for social bonding has made it possible for cats to live happily with people):

> Urban cats [can] live in large multi-male–multi-female groups reaching very high density. In these groups, cats exhibit frequent amicable interactions and olfactory recognition of group members versus strangers, while females cooperate in

rearing offspring. The social structure of males is organized around a linear dominance hierarchy. There is general agreement that the cat social system is a recent construct brought about by the changing environment due to human influence (dispersion of resources), and by the capacity of domestic cats to be opportunistic.[17]

Sex is big trouble in the country, but in the urban colonies, again, getting along seems to be more highly valued than individual advantage:

In rural areas males seem to be polygynous, fighting aggressively to copulate with as many females as possible, whereas in urban populations cats are promiscuous—males and females copulate with several partners. Aggressive behaviour among males during the breeding season is not severe, they allow subordinates to remain within the social group and reproduce.[18]

That study, obviously, was looking at cats with full sexual capacity. The denizens of Rome's shelters are all neutered, which makes them more peaceful yet.

Great, great. If only all feral cat life were like that. It isn't.

Conditions for ferals in the United States range from not bad to abysmal. Estimates of the national population vary by tens of millions. Definitions of "feral" vary, too. "Homeless" is one. "Un-owned" is another. "Semi-owned" cats are an especially saddening population because many of them are the unintended consequence of kindness.

One night after a literary event in the American South, I was sitting in a grand old parlor with a group of the gathering's rich, refined, and intelligent participants when the subject of cats arose.

A striking and sophisticated couple told me about a cat whom they believed they were helping. It (as they referred to the cat; they had not determined his or her sex) had turned up in their carport one evening, meowing. They fed it; it ate. They approached it; it ran away. It returned; they fed it again. Rain came, and winter. They put out an old towel and the cat slept there. This had been going on for six months.

Had they taken the cat inside?

—Oh, no. It wanted to come in, but they didn't want a pet. They traveled a lot.

Who fed it when they were away?

—It was just a stray.

Had they touched it?

—They thought it might have fleas.

Had they called the city, or the SPCA?

—Well, no.

Were they aware that some cats have a bar-coded microchip implanted in their backs, so that if they're lost they can be identified and returned to their owners?

—Well, yes, I suppose we have heard of that.

Had it occurred to them that some child six blocks away might have been crying her eyes out over her precious lost kitty? (I was getting mad.)

—No, there's this gully full of weeds down below the house, and all these cats...

Does this one seem wild?

—Not really, no.

Has it crossed your mind that it keeps coming back simply because you're feeding it?

—Yes, exactly. It might starve otherwise.

Or go home.

—We hadn't really thought of that.

This cat was an archetypal victim of semi-ownership. There are millions more. Sometimes people let them in and they sit on laps while the family watches TV till it's time to turn them back outside. Sometimes they're actually owned cats who keep getting fatter and fatter because "helpful" neighbors are feeding them— then suddenly, mysteriously, they slim down when the neighbors go on vacation. More often they're like that miserable, disoriented cat being intermittently fed in the literary sophisticates' carport, deprived of human company, in fact untouched, barely sheltered, fated never to get medical care again, bereft of any emotional connection.

This is the classic, almost universal behavior of the semi-owner— only to feed cats, and to do nothing further for them, least of all to have them neutered. The result, metastasizing nearly everywhere at this moment, is more and more free-roaming cats with nobody to look out for them.

There are many studies of the phenomenon. Here is part of one, from Alachua County, Florida, home of the University of Florida:

> Although provision of food for un-owned free-roaming cats was a common activity in the present study, few individuals who fed such cats took further action to have the cats neu-tered....At this geographic latitude, free-roaming cats may produce more than one litter of kittens each year. Given the high percentage of pet cats that were neutered, un-owned free-roaming cats likely represented the greatest source of cat overpopulation in the area. During 1999, 74 percent of the 4,827 cats admitted to the county animal control facility were classi-fied as strays, and almost half were kittens. A total of 3,714 (77 percent) cats admitted to the facility were euthanized....
>
> Many individuals do not covertly harbor these animals or feed them in secrecy....Individuals who feed un-owned

cats feel a protective bond for the cats they care for, even if the cats are too feral to be handled. Individuals who feed un-owned cats reported they would resist efforts to control the population of these cats by lethal means, and...most Trap-Neuter-Return programs are operated by private organizations capable of neutering, at most, a few hundred cats each year....The magnitude of the un-owned free-roaming cat population is much larger than the capacity of many existing programs....[19]

In Australia, over the course of the last two centuries, domestic cats have been significant contributors to the extinction of at least twenty-seven endemic mammal species. Australia's estimated twenty million feral and semi-owned cats prey on more than a hundred of the continent's currently threatened species. The situation is so bad that in July 2015 the government announced a program to kill at least two million feral cats.[20] One hardly need say that there was a furious public backlash. The French movie star Brigitte Bardot and the English singer-songwriter Morrissey have made the "saving" of Australia's feral cats an international cause, and the battle continues.

Meanwhile, a lot of Australians continue to feed free-roaming cats without the least thought of neutering them or providing any other kind of care (much less a thought of the harm they do). In a study of the situation, Australian researchers wrote, "Whilst these actions [feeding] [are] undertaken with a degree of benevolence in mind, they [are] unlikely to produce a favorable welfare outcome....Cat semi-ownership may be the product of cats being undervalued...regarded as disposable and able to fend for themselves."[21]

A piece in the *Dallas Morning News* of April 25, 2010, reported on an all too typical reader who "began feeding three kittens several years ago and now has a colony of about twenty cats. 'I have a real bad problem with feral cats and I brought some of it on myself, but I need some help [she wrote]. I do not want them killed and I don't know what to do.... There is nothing I can do because I'm crippled....'" The reporter contacted a nonprofit called the Feral Friends Community Cat Alliance. "The first time Feral Friends went trapping for [her]," he wrote, "volunteers caught four males and one female. They also found another neighbor who was feeding the rest of the animals and who promised to help with them. The group has more trapping rounds scheduled."[22]

In the United States, between 9 and 15 percent of all households feed feral cats, and very few of them take any further responsibility.[23] *Oh, you know, they're cats, they can take care of themselves....*

Elizabeth and I found ourselves confronting the problem of semi-owned cats up close. We had thought that our little postage stamp of a back yard in San Francisco was certainly going to be safer and more peaceful than the wilds of Montana, but we were wrong. The source of difficulty was the house next door, with which we shared the yard. The old woman who headed the household had lived there for decades, and she was very poor. The house was a twin to ours, built in 1875 and lavishly ornamented in San Francisco's famous "Painted Lady" style, but it had not seen paint in many years. There were holes in the plaster, and the floors were filthy raw wood. Soon after we moved in, I counted nine broken windows. Family members came and went irregularly. Sometimes we would hear furious shouting. In reply to our friendly greetings the old lady could often do no more than mumble through her frown.

The basement door hung open on one hinge, and behind it lay a jumble of broken furniture, bicycles, rotting cardboard boxes, stench—a prowling paradise for Augusta. As our first summer there ripened into warm October, Augusta would burst through her little flap-open cat door into the kitchen and drop on the floor the tiniest rodents I'd ever seen, so new, so hairless they seemed almost transparent, so nearly boneless that one chomp and a swallow were all it took her to ingest one. I chased one and grabbed it with a pair of tongs and examined it up close. It was a neonate Norway rat. Oh, nice.

We did not think little black Augusta a sufficient system of defense against a soon-to-be-burgeoning population of rats. But at almost the same time, a kitten about three months old appeared in the window next door. As she grew older, she was often to be seen in the window pawing against the glass, over and over, for an hour and more, as if to say, *Let me out, let me out!* The people there did feed her, but for reasons we never knew they did not touch her, and she was, in effect, semi-feral. One of the broken windows, giving onto the back yard, was close enough to the ground that it could have offered egress and ingress to the cat, but it had been blocked by a piece of cardboard. Then one day the cardboard was withdrawn. Thereafter, in the evenings, we would hear one of the daughters of the house calling the cat by name, over and over, "Devon, Devon, Devonnn." Sometimes Devon came for a furtive meal, sometimes she would be gone for days. Clearly she had found companions in the neighborhood. The nights were filled with the squalling, hissing, and shrieking of cats.

About nine weeks later, Devon came home and gave birth to a litter of three kittens. "I don't know what to do," the daughter said. Consequently, neither she nor anyone else in the house did anything. Well, no, they did feed Devon, and in due course the kittens, irregularly. Also they left the broken window unblocked. Perhaps Devon hunted late at night.

One day I saw one of the kittens, maybe ten weeks old, squashed flat in the middle of our busy street. Over the course of the next twenty-four hours no one came to remove the body. I shoveled it into a plastic bag and dropped it in a public trash can on the corner.

About this time I discovered that somewhere down the block we had our own version of a Roman *gattara*, who put out enormous amounts of dry cat food late at night for—well, for come who may. Her several feeding stations in the vicinity attracted dozens of feral cats, as well as raccoons, crows, ravens, and rats who horned in on the surplus. It didn't take long for Devon and her two surviving kittens to discover these nocturnal feasts.

I couldn't bear the idea of keeping our freedom-loving wild-at-heart Montana cat confined to the house. Augusta could bound to the top of our six-foot-high fence with ease—and disappear. We had a shaky confidence that our entire block—of houses, apartment buildings, and little shops all contiguous, on all four sides—afforded no access to the street. (Then how had Devon's kitten gotten out to be run over? Another broken window, we told ourselves.) In any case Augusta had always had a deathly fear of the sound of internal combustion engines. As for the dangers of feral gangster cats, it was a rule that Augusta must never be allowed out after dark, for that was when the bad cats roamed. If she was out in the late afternoon, we called her and she would come.

A year later, after Devon's first two kittens had drifted off into the dark universe of ferals, she had another litter—a girl named Mocha, a boy named Snowy, and another male whom I called Feraldo. Like their mother, they were in and out of the house through the broken window and occasionally fed, but they were even wilder and more untouchable than their mother. Mocha and Snowy stayed close to home, however, while "Feraldo" disappeared.

By then Augusta could no longer leap to the top of the fence: She had hip dysplasia, a rare condition in cats—rare enough not even to have an entry in the comprehensive *Cornell Book of Cats*. It was thought to be usually an affliction of purebreds, especially heavy-boned ones like Persians and Maine coons;[24] Augusta was notably fine-boned, and the furthest thing from purebred. What we did not know, and at that time our vet may also not have known, was how common degenerative joint disease actually is in cats: Osteoarthritis may have been what caused her hip dysplasia, rather than the other way around. It was not until the early 2000s that researchers found that between 40 and even 90 percent of all domestic cats show some symptoms of joint deterioration.[25]

Though no longer able to bat a hummingbird out of the air, Augusta seemed content to watch birds at the feeder. Then in the summer, in Montana, freedom resurrected her strength: She could climb trees, bound as ever in great arcs through tall grass, dash at a fleeing vole, pounce on it, seize it, and bring it proudly home. No, Augusta, you may not come in with that thing in your mouth, you have to kill it outside.

Within less than a year, Devon was pregnant again, and Elizabeth and I said to each other, This has got to stop. We went to our next-door neighbors and said, Listen, you know Snowy and Mocha are old enough to have kittens of their own. We've called the city, and they told us that the SPCA will neuter feral cats for free, and they'll lend us Havahart traps, which don't hurt the cats at all. We'll catch them, and we'll take them, and they'll have their little operations, and we'll bring them back. Okay?

The neighbors meekly acceded.

We got the traps, we caught the cats, and we did what we said we were going to do. Snowy and Mocha each had an ear clipped short, just like Roman cats, to signify their new non-reproductive status. When the kittens were old enough to be fostered—four weeks—we went to the neighbors again, and asked if they would let us take Devon to be neutered, and let the SPCA give the kittens to a foster home, where they would be raised to be tame, and when they were ready, they would be adopted into good homes (unsaid: not be run over in the street or disappear into oblivion). This was harder for them, but again they agreed.

Devon gradually became almost a house cat. Though she would let no one else touch her, she was able to bring herself to sit in the lap of the daughter of the house. Snowy and Mocha continued to lurk around the neighborhood, feeding mostly at the cat lady's heaps of kibble in a church parking lot and fleeing at the sight of any other human being. At some point, both of them were trapped, and Mocha was evidently tractable enough to be adopted. Snowy was too wild, however, and so was spayed and returned. Years later, when the old lady died and the daughter sold the house and moved away, she took Devon with her and left Snowy to his own devices. As of this writing, he is still around, filthy and rather fat. The cat lady's feeding continues.

Snowy's life is better by a considerable degree than those of most other feral cats in America. The San Francisco Society for the Prevention of Cruelty to Animals has one of the most successful Trap, Neuter, and Return programs in the country. According to Cynthia Kopec, chief operating officer of the SFSPCA, in 2015 the "live release rate"—the fraction of cats taken in by the shelter and then either adopted, returned to their owners, or transferred to another nonprofit that will guarantee finding them homes—was 96 percent, the highest of any major American city. Since 1989 the SFSPCA has had a rigorous no-kill policy. The citywide rate of euthanizing cats and dogs is 0.0009 per capita. Only mortally wounded and fatally ill animals are euthanized. About a thousand feral cats are trapped, neutered, and returned to their colonies every year, and for the last two years all the known colonies are tracked on the SPCA's computer system. The apparent total population of feral cats in the city is trending down. San Francisco is lucky in being surrounded by water on three sides, so that in-migration from adjacent feral cat populations is limited. Until recently, the city's thousand-acre Golden Gate Park was swarming with feral cats, and they were blamed for the virtual eradication from the park of its entire population of about 1,500 California quail—the California state bird. The feral cat population there is now down to about thirteen. The quail have yet to return.

San Francisco is rich, and soft-hearted. The SPCA's success there owes a lot to those financial and cultural advantages. Elsewhere in the United States, there are valiant efforts in nearly every major city to educate the public about the need to neuter their pets. There are nonprofit organizations—local SPCAs and many others—that lend traps and provide free neutering by volunteer veterinarians. Some cities have adopted total no-kill policies and have established expensive, municipally funded TNR programs. But nearly everywhere it is a losing battle.

According to Dallas Animal Services, that city "has approximately 350,000 homeless, wild, or untamed cats," and it's generally agreed that that population is growing. The agency now has an active Trap, Neuter, and Return program and provides "colony managers" for each targeted population. Euthanasia is sharply down from the rate of a decade ago, but the city still kills more than ten thousand cats a year.[26]

In Atlanta the nonprofit group Catlanta has sterilized more than thirty thousand feral cats, and has enabled the neutering of another thirty thousand by independent veterinarians. But an additional twenty-five thousand feral cats end up euthanized in Atlanta's shelters every year, at an annual cost to the city of $3.5 million.[27]

Baltimore has about 185,000 feral cats.[28] Chicago, half a million.[29] Los Angeles, three million.[30] The Humane Society of the United States estimates the national population of feral cats to be thirty to forty million.[31] The American Society for the Prevention of Cruelty to Animals says, "It is impossible to determine how many stray dogs and cats live in the United States; estimates for cats alone range up to seventy million."[32] The *PBS News Hour*, not citing a source, reports "an estimated eighty million."[33] *New York Times* reporter Bruce Barcott reports "fifty to ninety million," but then adds, "(exact figures are impossible to ascertain)."[34] Without estimating an ongoing total, Becky Robinson, president of Alley Cat Allies, looks at reproduction, stating, "Forty million feral kittens will be born throughout the country this year, but twenty million of them will die at birth."[35]

Another way of considering the magnitude of the feral cat population is to measure its effects. The most comprehensive study, a study of studies, from the Smithsonian Conservation Biology Institute, published in 2013, asserted: "We estimate that free-ranging domestic cats kill 1.4 to 3.7 billion birds and 6.9 to 20.7 billion mammals annually. Un-owned cats, as opposed to owned pets, cause the majority of this mortality. Our findings suggest that free-ranging cats cause substantially greater wildlife mortality than previously thought and are likely the single greatest source of anthropogenic mortality for U.S. birds and mammals."[36]

This was big news. In the *New York Times*, Natalie Angier wrote that the Smithsonian study was "the first serious estimate of just how much wildlife America's vast population of free-roaming domestic cats manages to kill each year.... The estimated kill rates are two to four times higher than mortality figures previously bandied about, and position the domestic cat as one of the single greatest human-linked threats to wildlife in the nation."[37] The same primary researcher, Scott R. Loss, and his co-authors also looked at bird mortality caused by collisions. The range of the estimates was very wide—89 to 340 million birds per year killed by motor vehicles[38] and 365 to 988 million by buildings[39]—but, even at the high end, those numbers pale beside the bird deaths attributable to feral cats. It's important to note here that the Smithsonian study states up front that "Un-owned cats, as opposed to domestic pets, cause the majority of this mortality."

The study sounds alarming, but there's a great deal it doesn't tell us that might make the numbers less frightful. It doesn't say which bird species are most affected, and many species are capable of rebounding from intense predation. If we're going to control the truly harmful predation by cats, we need to know where to concentrate our efforts. There is also the question of habitat loss, which is irreversible and which is the prime factor in the decline of rare species of not only birds but other animals as well as plants all over the world. In short, the lack of specificity in the Smithsonian paper makes it a good spur to more accurate research, but is hardly cause for panic.

In the United Kingdom, no less fierce a gang of bird lovers than the Royal Society for the Protection of Birds accepts that cats kill about 55 million birds a year, but they nevertheless state: "There is no scientific evidence that predation by cats in gardens is having any impact on bird populations UK-wide.... Those bird species that have undergone the most serious population decline in the UK (such as skylarks, tree sparrows, and corn buntings) rarely encounter cats, so cats cannot be the cause of their declines. Research shows that these declines are usually caused by habitat change or loss."[40]

There is a true global emergency, however, on islands, much of it attributable to feral cats. Because of their isolation, most island-dwelling animals never evolved in the presence of mammalian predators and therefore have little or no defense against them. Island populations are particularly vulnerable to extinction, because when they are reduced to very small numbers there are usually no adjacent populations to restore those losses; and very small populations of any living creature have a very high risk of extinction simply owing to random fluctuation of environmental conditions and their own reproductive success or failure (failure often abetted by inbreeding). Domestic cats have spread to most of the planet's islands, and because the species is adaptable to so many environments, feral populations have thrived on many islands.

Feral cats' prey can be identified by analysis of their stomach contents or scat. Using that technique, a meta-analysis of seventy-two studies of feral cats' diets showed that "at least 248 species were preyed on by feral cats on 40 worldwide islands (27 mammals, 113 birds, 34 reptiles, 3 amphibians, 2 fish, and 69 invertebrates)....At least 175 vertebrate taxa (25 reptiles, 123 birds, and 27 mammals) are threatened by or [have been] driven to extinction by feral cats on at least 120 islands."[41]

The International Union for the Conservation of Nature has documented the extinction of 238 vertebrate species; feral cats contributed to 14 percent of those extinctions. Feral cats threaten the survival of 8 percent of the 464 species now listed by IUCN as "critically endangered."[42]

Cats don't even have to kill birds to reduce their populations. Birds simply *scared* by cats sometimes don't reproduce at their normal rate. A study at the University of Sheffield in the United Kingdom shows that "a small reduction in fecundity due to sub-lethal effects"—namely, fear—"can result in marked decreases in bird abundances (up to 95 percent). Thus, low predation rates in urban areas do not necessarily equate with a correspondingly low impact of cats on birds. Sub-lethal effects may depress bird populations to such an extent that low predation rates simply reflect low prey numbers."[43]

The diseases you can contract from feral cats are really scary. Feral cats are now the primary source of rabies in the United States—60 percent of all cases. Parasites in feral cat feces, which can be easily picked up in playgrounds, sandboxes, gardens, and the like, can cause nerve damage, blindness, abortions, and birth defects, particularly hydrocephalus. Hookworms attack through bare feet on beaches and lawns, and cause skin lesions, pneumonitis, muscle infection, and eye disease. Cat scratch disease produces fever, headaches, and lymph node enlargement; 5 to 15 percent of cases involve encephalitis, retinitis, or endocarditis, which can be deadly. Eight percent of plague cases in the United States originate in contact with cat fleas. Worse, cats "frequently develop the pneumonic form of plague, which is considerably more infectious to humans…and results in rapidly progressive and frequently fatal disease."[44] There's more, but that feels like enough.

There lately has been a flurry of alarmist baloney about a protozoan parasite common in cat feces called *Toxoplasma gondii*. It is in fact very easily transmitted to humans and other animals—about a third of all people alive today are carrying it. The alarms have been ringing about toxoplasmosis causing (or being "correlated with"—always a sign of shaky science) schizophrenia, major depression, suicide, criminality, memory loss, mental decline, poor impulse control, road rage, and more. It does seem to be bad news for some marine mammals, and immunocompromised people can be badly affected too. But after a string of years with the sky about to fall, a team of researchers led by the neuroscientist Karen Sugden of Duke University found, quite definitively, "little evidence that *T. gondii* was related to increased risk of psychiatric disorder, poor impulse control, personality aberrations or neurocognitive impairment."[45]

It's easy to think of feral cats as agents of harm. What we must remember is how they suffer, how miserable their lives are. Feral adult females average 1.4 litters a year, with a median litter size of three, but 75 percent of the kittens die or disappear before the age of six months. The median survival time for feral kittens is 113 days.[46]

Feral cats don't get old. People poison them, torture them, and shoot them. They die from diseases and infections that a veterinarian could easily cure. Often they're run over and then die of wounds that with medical care could have healed.[47] They're captured and sold for illegal medical experimentation. Dog fighters use them as bait.[48]

There seems to be universal agreement that there are too many feral cats—that wherever feral cats occur, in fact, there are almost always too many of them. There is not agreement, however, about whether certain feral cat populations should be reduced to smaller sizes or reduced to zero, nor is there agreement on the question of whether the world would be a better place if there were no feral cats at all. Two related questions, much debated, are, one, Can feral cats have a good life? and two, Owing to all the damage they do, as well as the suffering of the cats themselves, is the feral life by definition not a good one?

Population reduction—the proposition on which we seem to have unanimous agreement—would address both those issues: Harm to other species would be reduced, and the welfare of the surviving feral cats would increase. So far, so good. Here, however, our proposition hits a brick wall. The disagreement over how to reduce feral cat populations has become a virtual war. Let's call the antagonists the Cat Faction and the Bird Faction.

The Cat Faction puts the welfare of feral cats first. They tend to deprecate the impacts of predation by feral cats—in fact many of them refuse to face the facts at all—and they believe in Trap, Neuter, and Return. They believe, in fact, that properly conducted, TNR is capable of reducing feral cat populations to zero—if that's the goal agreed to.

Theoretically, the Cat Faction's case for TNR is unassailable. As the trapping and neutering succeed, the number of reproducing cats falls. They can point to TNR programs all over the world

that at this moment are bringing about declines in feral populations. SPCAs across America have charts and graphs documenting
their success. But there's a problem. To re-quote Eugenia Natoli:
"TNR programs alone are not sufficient for managing urban feral
cat demography."[49]

Eighty percent of the eighty-million-odd owned cats in the
United States are neutered, which means, ipso facto, that 20 percent
are not. When you break down the 80 percent by family income
grouping, you start to see where the problem is concentrated.
Ninety-six percent of cat-owning families with a household income
of more than $75,000 have had their cats neutered. For families with
an income between $35,000 and $75,000, the percentage with neutered cats falls only a little, to 91. But of households with an income
below $35,000 only 51 percent have had their cats neutered.[50]

Most cities and many rural areas offer neutering at low to no
cost. It may be that more than a few people living in households
with low income just don't know what's possible. That was certainly the case with our next-door neighbors in San Francisco. A
further problem is that lower-income people are disproportion-
ately likely to feed semi-owned neighborhood cats as well as ferals.
The San Francisco SPCA's computer map displays a pin for every
un-owned cat they've trapped in the last two years, and those pins
are most crowded in the city's least advantaged areas.

A study comparing rates of cat mortality in United States census
tracts in Boston found a strong correlation between low household
income and a high death rate for cats (both owned and un-owned).
Cat mortality in the poorest tracts was more than *forty times* that
in the richest. Not surprisingly, premature human death was a lot
more common in poor neighborhoods as well.[51]

One long-term study of a TNR program with a heavy emphasis
on adoption, on the campus of the University of Central Florida in
Orlando, found quite favorable results. The study and the program

began in 1991 with a population of 155 free-roaming cats—some socialized to people, some feral. It wasn't totally no-kill: Severely ill cats were euthanized, and there was intensive trapping of the rest. By 1996 the population was down to 68, and all but one, a tomcat, were neutered. New arrivals were not allowed. By the end of the study in 2002, only 23 cats remained in the colony. Half of the original 155 had been adopted. The quality of life of the survivors was greatly superior to that of the original population.[52] A study like this supports the Cat Faction's contention that TNR can work. It must be realized that a lot of volunteer labor is indispensable, and if abandoned cats and kittens make their way into the population, the program will fail.

The Bird Faction is led primarily by scientists and conservationists, most of whom will tell you that they love cats, and they often own cats themselves, but they see *feral* cats as essentially an invasive species. They also have a number of studies demonstrating that Trap, Neuter, and Return simply doesn't work. Some scientists have developed innovative alternatives that could work well in combination with other approaches.

The disagreements have been passionate. One report from the University of Nebraska—admittedly having originated as an undergraduate project but ultimately compiled by faculty members—proposed the use of "integrated pest management," suggesting habitat modification, fencing, netting, chemical repellents (among them benzyldiethyl [2, 6-xylyl carbamoyl methyl] ammonium saccharide, "the world's bitterest known substance"[53]), and "shotguns with No. 6 shot or larger, .22 caliber rifles, or air rifles capable of shooting 700 feet per second or faster." Also Trap-Neuter-Vaccinate-Release.[54] Shooting a cat, by the way, is illegal in all fifty states.

Needless to say, that proposal didn't get much traction, but most serious scientists ultimately conclude that humane euthanasia is a necessary part of any feral cat population reduction plan, however repugnant the idea. Quite a bit of thought has gone into possible ways of resolving the conflict, but this paper aptly sums up the difficulty:

> We conducted a survey of opinions about outdoor cats and their management with two contrasting stakeholder groups, cat colony caretakers (CCCs) and bird conservation professionals (BCPs) across the United States. Group opinions were polarized, for both normative statements (CCCs supported treating feral cats as protected wildlife and using Trap-Neuter-Release (TNR) and BCPs supported treating feral cats as pests and using euthanasia)....Most CCCs held false beliefs about the impacts of feral cats on wildlife and the impacts of TNR (e.g., [only] 9 percent believed feral cats harmed bird populations).[55]

Except for the rare cases in which abundant resources, intensive effort, and colony isolation combine, sooner or later the Cat Faction is going to have to admit that TNR is not, alone, the perfect solution. The American Veterinary Medicine Association wrestled for years with the problem, and finally, at its meeting in February 2016, after long, exhausting argument, for the first time the AVMA's policy on free-roaming cats made this assertion—with the kicker in the last sentence:

> The AVMA recognizes that managed colonies are controversial....The goal of colony management should be continual reduction and eventual elimination of the colony through attrition. Appropriately managed colonies also have the potential to significantly decrease risk to public health, wildlife, and ecosystems. For colonies not achieving

attrition and posing active threats to the area in which they are residing, the AVMA does not oppose the consideration of euthanasia when conducted by qualified personnel, using appropriate humane methods.[56]

Here's the thing. The temperature on this could be turned way down if one big problem were addressed: abandonment. As Eugenia Natoli says, it has to start with education. But what about a little more pressure? What about penalties? Cruelty to animals is illegal in every state in the United States, but each state has its own definition, and abandonment isn't always included. It isn't rare for people to try to give up a pet to a shelter, sometimes for a pretty lame reason, and have the shelter just say no. The San Francisco SPCA does its best to talk people with healthy pets into keeping them, and they try to help, even financially, with animals that are sick or have behavior problems—they even have a full-time behaviorist on staff.

Other shelters have also been developing pet retention programs. When people can't afford the fee to reclaim a lost cat, or the price of medical care, or even cat food, they often come in tears to a shelter, at their wits' end, to give up a beloved cat. These new programs can provide financial aid, food, and free veterinary care, including neutering and microchipping. They will even help people find cat-friendly housing and pay the pet deposit on a new apartment. Sometimes there are temporary foster homes that will care for a cat to tide the owner over through a rough patch—a day, a week, even months. This kind of thing saves cats' lives.

Where assistance like that is not available, and shelters won't accept a surrendered cat—sometimes because they sense that the person's stated reason for giving up the cat is a lie—what happens next is usually abandonment. Which in turn most likely means a very bad death—the cat run over or starved. Whoever dumped Augusta in the Montana snow may well have said to themselves, "Oh, look at the smoke coming out of the chimney, what a nice place, I just know they'll take her in." Which of course we did. But I've always believed she had littermates who died that same night.

In any case, the Bird Faction had better give up the idea that just rolling out the data is going to change the minds of the Cat Faction. It's now well accepted in psychology—and increasingly evident in politics—that "Head-on attempts to persuade can sometimes trigger a backfire effect, where people not only fail to change their minds when confronted with the facts—they may hold their wrong views more tenaciously than ever."[57]

And yet, on the other hand, one of the most tenacious, indeed ferocious, animal rights organizations, People for the Ethical Treatment of Animals, now says, "We believe that it can be marginally acceptable to trap, vaccinate, alter, and release feral cats when the cats are isolated from roads, people, and animals who could harm them, are regularly attended to by people who not only feed them but also provide them with veterinary care, and are kept in areas where they do not have access to wildlife, and the weather is temperate. The biggest problem is that most cats, once they are caught to be sterilized, will not be able to be lured by traps again when they are sick or injured.[58] A painless injection is far kinder than any fate that feral cats will meet if left to survive on their own."[59]

So PETA and America's veterinarians are softening. Meanwhile the non-softening antagonists may wish to find some things that most of them can agree on. There have been some imaginative alternatives proposed. The Birdsbesafe collar is a big, vaguely Elizabethan, brightly colored thing that perhaps a few cat owners with little concern for esthetics might try. Apparently it scares the dickens out of any bird the cat comes near.[60]

Cat depredation—indeed cat presence—is greatly reduced in any area where coyotes are active.[61] Those are generally less densely settled places, but often rich in bird life. Density is beginning to matter less, however. Coyotes have been expanding their range in the United States for the last fifty years, and they are more and more often seen in suburban and urban settings. When coyotes arrived in the Presidio of San Francisco (possibly from coyote-rich Marin County by crossing the Golden Gate Bridge!), the abundant feral cat population evaporated quite quickly.[62] The coyotes may have killed a few cats, but more significant, probably, was that when a cat saw a coyote, the cat got out of Dodge. They just knew. Coyotes will in fact readily kill cats, not necessarily to eat them—perhaps the coyotes think of the cats as competition, for cats' prey base of small animals overlaps quite a bit with coyotes' rather broader one (coyotes are not exclusively carnivorous). Cat owners in areas frequented by coyotes must learn to keep their cats indoors, at least from dusk to dawn.

The only hope for feral cat colonies that people are determined to protect in coyote habitat is the construction of well-fortified sanctuaries. Sanctuaries, in fact, may be an excellent solution for feral cats in all sorts of places. Proper construction would prevent not only depredation on wildlife by roaming cats but also the surreptitious introduction of abandoned cats. Life inside the sanctuary could be well regulated and peaceful. Once a successful TNR program had reduced a feral population to a reasonable size, this could be a long-term, permanent solution, and euthanasia would need never to have entered the picture.

Alley Cat Allies fiercely oppose the idea, but their conception of the sanctuary seems to be a small, indoorsy place jam-packed with cats, riddled with disease, and still surrounded by thousands of unprotected feral cats.[63] How about nice big rural places? With populations of *reasonable* size? I fear that too many in the Cat

Faction may believe that maintaining the current vast overpopulation sounds like a desirable norm. It isn't. It's not desirable and it's not possible.

How about licenses? Dogs have to have licenses, why not cats? Think what municipalities could do with the money—think how many free sterilizations they could perform or pay for. Alley Cat Allies are against that, too, contending that every cat seen without a collar would be taken to a shelter and probably euthanized.[64] Really now, that wouldn't have to happen. Maybe the license takes the form of the microchip that all thoughtful cat owners already are implanting just under the skin of their kitties' backs, with full identification in digital form, which can be read like the bar code on your box of Cheerios. It's painless, it's cheap, and it's as close to error-free as any ID system in the world. Alley Cat Allies say only 2 percent of cats in shelters are ever reunited with their owners. Put microchips in 90 percent of a town's cats, and 90 percent of lost cats will go back home.

Here is a proposal. This would take a while, but it is feasible. Let the states pass laws mandating the licensing of all cats, using implanted microchips. The licensing fee must be very small—perhaps free if you can't afford it. Every person who takes a cat to be neutered gets a cash payment of one hundred dollars (and a license if the cat doesn't have one). If the person declines the money, they get a one-hundred-dollar tax deduction. The money comes from private groups and government grants. It will not be long before governments realize they are spending less on that program than they previously spent rounding up and sheltering stray cats.

The average price charged by veterinarians to implant a microchip is about forty-five dollars; that includes registration in the pet-recovery database.[65] A portable scanner—which every animal control officer would carry—costs less than a hundred. A veterinarian typically charges between one and two hundred dollars to spay a female cat, fifty to a hundred to neuter a male.[66] Although there are plenty of programs that offer discounted or free chip implantation and sterilization, public awareness of those resources is woefully short. Mandatory licensing, with subsidy for any cat owner who couldn't afford the chip and the neutering, would solve that problem.

If the animal welfare groups—ASPCA, the Humane Society, PETA, maybe the National Audubon Society (which is full of bird lovers)—could get together on raising a cooperative fund and backing consistent legislative proposals for state licensing laws, then the metastatic overpopulation of feral cats could be drastically reduced in a relatively short time with little or even no euthanasia involved. I know, this starts to sound like one of these if-everyone-would-light-just-one-little-candle pipe dreams. But what we have now is not working, so maybe some dreaming isn't a bad idea.

What Rome does will never happen in the United States, nor, probably, anywhere else. Even Rome's success is limited, in fact, and, once you leave the city, Italy's feral cats lead characteristically miserable lives, which often end violently. San Francisco seems to have the money and the determination to succeed in a 100-percent no-kill approach to its feral cats, but so far it is in fact not succeeding. The SFSPCA will tell you that the city is making progress, that the feral cat population will eventually dwindle near zero. That may be true; we will see. But as in Italy, for every almost-ideal example of successful TNR, clean shelters, and plenty of volunteers, there are hundreds of instances of failure, feral populations completely out of control, no funding for humane programs, and no prospect of it. Where else in the world are there counterparts to Rome's chic and generous *gattare*?

We who have our happy, healthy cats at home, our Augustas, and through them love all their kind—what shall we do?

Chapter Six

A Good Life

The little log house was so cowboy-chic it had been on the cover of a Western décor magazine—the Navajo rugs, the sepia Edward Curtis photographs of tragic chiefs, the furniture with bark still on the wood—but it sat in a seething nest of rattlesnakes. The stair-step cliff that rose behind the house was in effect a solar oven, down which, that summer of drought, the rodents of the high prairie above had come in quest of water, pursued by a multitude of vipers. The snakes liked the heat. They basked easily seen in the cliff's declivities and lurked invisible in the tall, dry, ungrazed grass between our front steps and the river. The friend from whom we had rented the house had hidden the key between the third and fourth of the stack of round flat stones beside the door, and when I lifted the top three, there lay curled a tiny rattlesnake, which, the moment light hit it, raised its little tail to rattle and its little head to strike. But when the evening chill came, and it came quickly, torpor quieted the snakes all around, and Augusta was (I told myself) safe.

It was her second summer there, familiar ground, but unsteadier ground than the summer before, owing not only to the drought and the rattlesnakes but also to domestic non-tranquility, which registered in her anxious eyes and an extra measure of emotional need. This was the summer when we returned from an evening out to find her far from the house in the middle of the bridge above a furious passage of the West Boulder, disoriented and scared, possibly having been trying to follow us, possibly even seeking her kittenhood home five miles upstream. After long work on a comic novel set in Montana, I had sent it to my longtime literary agent— and she had hated it. New agent after agent had been rejecting me. Now at last the rising star David McCormick and I had found each other, and our first meeting, in New York, was imminent. My flight and my room at the Yale Club were reserved, but it was not a good time to be leaving. My father suffered from severe macular degeneration—he was going blind—and I had finally gotten him an appointment at the Wilmer Eye Institute at Johns Hopkins, to which he was incapable of going without me. But it was not a good time to be leaving. The heat in the house, physical and emotional, was oppressive, but it was not a good time to be leaving. Augusta chose this time to disappear.

We looked everywhere, we called and called, we alerted the sheriff, the veterinarians in equidistant Livingston and Big Timber, every neighbor up and down the West Boulder valley. Two mountain-climbing friends scaled the cliff to a golden eagle aerie— a dozen feet across—set into a shallow cavity and deep with the tattered remains of their prey. The biologist studying eagles in the Boulder drainage told me there could be a century's worth of bones in such a nest. My friends did not find pieces of Augusta. I heard about a mountain-lion-hunting guide who trained his hounds on domestic cats. They could sniff Augusta's bedding and track her to wherever she was. I booked him. They were set to go the very next day. But early that morning, before the sun rose, it snowed—Montana in the middle of June!—obliterating all trace of scent.

Augusta, Augusta, *Auuugussstaaa!* we called, day and night, and then it was time—there could not have been a worse time—for me to leave. Like so many times when I did wrong by Augusta, and more times, too, unrelated to her altogether, when I was simply bound and determined to do some damned thing for reasons I can no longer reconstruct, I could not change my mind, and I should have changed my mind. My new agent would have understood. I could have gotten another appointment at Johns Hopkins for my father. The airfare change fee, duh, I could have paid that. Augusta, what could have possessed me? I think I knew about as much as you did: Something was wrong and I ran away.

I went to New York, I went to Baltimore, I went to a friend in the Berkshires and wept for her sympathy. I talked to Elizabeth every day on the phone, and although she was, as she recalls it, "spending hours searching, calling, crying, worrying, lying awake, I might have talked through every detail with you less and less." Which gave me the impression that she was doing less and less, caring less and less. I realize now that that was a delusion, rooted in a briar patch of self-misunderstanding. I was not in my right mind. If I, *I*, cared so much, what in God's name was I doing two thousand miles away in Massachusetts?

Augusta grew up outside in Montana, okay? She played with bears. She escaped from coyotes. She knew that that culvert she took shelter in was too small for them. Smart kitty! Loud city noises and big men aside—scary things she had not been habituated to as a kitten—she was no fraidy-cat. I had many and elaborate justifications for Augusta's outdoor freedom. You might have argued that the life I allowed Augusta in both San Francisco and Montana was an egotistical folly, and I could have argued you blue that it wasn't.

I could argue on more than Augusta's individual behalf. Having felt in myself—even if only in imaginative projection—the joy that rippled through Augusta's frame when she engaged the wild wind and the countless immingling scents it bore, I was certain that that joy was at least potential in all her species, and certain that too many cats never know it. Sorting through that complexity, making sense of it—that was how she formed her world. A certain *turbulence*, not only of scent but of experiences, too, is resolved, set in order, in the cat's tranquility. That is her genius; that is the richness of her life. Complexity is a condition indispensable to a cat's good life.

The setting doesn't have to be some pristine redoubt, far from it. A city park is thrillingly alive with olfactory fascination for cats; but they must have learned to tolerate the resistances it presents to a delicately nurtured feline sensibility—children, car horns, barking or over-curious dogs, and the like—and acceptance of a sturdy harness is good as well. They can learn all that easily, if they're started early and gently enough. An additional benefit, for the person who has made the effort, will be a companion of notable equanimity.

Real country cats may require no constraint at all as long as you accept the risks and exercise reasonable caution. But please— not if you're close to a road. Automotive traffic is the greatest danger anywhere, and although most cats are terrified of the sound of motor vehicles, the evidence of their inconsistent judgment is to be seen on roadsides everywhere, even backwoods routes seldom traveled. Some cats can be trained to stay away from the road, some will just naturally do it, and then it's a chance you can decide to take or not. After that decision, there are still plenty of dangers. Almost anywhere in the American countryside, you probably have coyotes in your neighborhood, though you may not know it, and they surely do kill cats. There are all sorts of other animals that will kill a country cat, some of them pretty small, including feral cats, especially intact males. Some of the killers drop out of the sky, where cats tend not to be looking—hawks, owls, gangs of crows for the sheer hell of it. Sometimes it's just canine amusement, *arf arf arf* and up a tree she goes. But not rarely, alas, either by instinct alone or even egged on by their masters, dogs kill cats.

The list of mortal perils is long, and by no means all external. One very thorough study of suburban cats found that risky behavior by far outweighed the risk of attack. The greatest risks, in descending order, are 1) the aforementioned danger of crossing a road; 2) exposure to infectious disease from other cats or from fleas or ticks, even the possibility of poison from common toads; 3) eating or drinking the wrong thing (antifreeze seems to be an especially evil temptation); 4) exploring drain systems, where they can drown in sudden storms; and 5) getting locked inside the crawl space of houses, there to die, horribly, of starvation.[1] Not surprisingly, just like teenage boys, male cats are much more likely to take stupid chances.

Not very often, but still worth thinking about, some people just shoot cats. The town we most often went to when we lived in Montana, and where I've spent two recent summers, Livingston, was home to a gentle-seeming citizen, a friendly neighbor, who by night shot cats down and then bludgeoned them to pulp with a shovel even when they were already dead. Jim Durfey was so confident of his impunity that he disposed of the remains in his own garbage bin, but he made the mistake of killing a cat that belonged to Bonnie Goodman, one of Livingston's leading cat lovers, who campaigned for his prosecution, and succeeded. Durfey was fined five hundred dollars, charged ninety dollars in court costs, and sentenced to a year in jail—deferred on the condition that he give up shooting cats for a year. When the judge asked him what he had learned from his experience, Durfey said he had learned that traps were available from the city.[2]

Ten years later, cats are still disappearing in Durfey's neighborhood, but there have been no reports of gunfire. Neighbors have seen a trap in his back yard, but that's all anybody is saying.

Jim Durfey wasn't the only guy in town with what you might call a cavalier view of the species. In the police log of the Livingston *Enterprise* newspaper of July 26, 2016, the following entry appeared:

> • 1:36 p.m.—A caller at the Park County Fairgrounds reported they saw a Park County department truck dragging a dead cat behind the vehicle. LPD [Livingston Police Department] and animal control responded and spoke with the county employee driver. The driver had been dispatched to dispose of a dead cat and, because it smelt foul, opted to drag the dead animal to dispose of it. No criminal action was taken, and the employee was advised to dispose of the animal in a different way.[3]

Well, at least that cat was already dead.

These things happen everywhere. In the spring of 2016, in London, the "Croydon Cat Ripper," under cover of night, had mutilated and killed scores of cats, including "Oscar, an eight-year-old tabby, his head and tail cut off; Oreo, a Siamese kitten left at the bottom of a garden, decapitated and with her collar lying across her body; and Paddy, who had been hacked in half."[4]

Montana's Jim Durfey made vague noises about saving birds from cat predation. The Croydon Cat Ripper had made no public statement.

It has been a sweet comfort to the owners of outdoor pet cats to learn that ferals bear most of the guilt for the slaughter of songbirds now under way, but few of us have not been presented with feathered gifts from our proud hunters. It is decidedly worth knowing, however, that predation by pet cats can be pretty spotty, depending on local conditions. Researchers from the University of Georgia and the National Geographic Society's remote imaging project monitored fifty-five suburban cats wearing miniature video cameras for periods of seven to ten days—for a full year, in order to account for seasonal variation—and got some interesting results. First of all, their owners way underestimated how many of them were hunters: 44 percent. But they weren't great hunters: For every week of hunting, they captured a grand total of 2.4 prey items. The cats brought fewer than a quarter of those home. They ate slightly more than a quarter, and the rest—more than half—they just killed and left on the ground. Obviously they were getting plenty to eat at home.

What they did kill would not, I think, stir the ire of a conservation biologist. Only sixteen of the fifty-five cats were successful hunters. More than a third of their victims were little reptiles—Carolina anoles, five-lined skinks, three snakes. They killed a butterfly, a walking stick, a dragonfly, an unidentified flying insect, a frog, three worms, four voles, one shrew, three chipmunks, one mouse, and a squirrel. The haul of birds was pretty poor—one robin, one hermit thrush, one eastern phoebe, and two unidentified nestlings.[5] The biodiversity of Clarke County, Georgia, was not significantly degraded.

Other research has also shown sharp differences in predation by feral cats and pet cats in the same area. In a study in central Illinois, owned cats had smaller home ranges than un-owned cats. Un-owned cats were more active at night, and because their prey was more available then, they were far more successful hunters; they were also much more likely to die young.[6] Owners of cats who let them roam can dig up plenty more such studies. But like everybody else in every other field of study, they tend to find what they want to find. One good, slightly sneaky study in England observed, "Cat owners generally disagreed with the statement that cats are harmful to wildlife, and disfavored all mitigation options apart from neutering." (Here we go again!) "These attitudes were uncorrelated with the predatory behavior of their cats. Cat owners failed to perceive the magnitude of their cats' impact on wildlife and were not influenced by ecological information."[7]

But! A study in Switzerland showed that only 16 percent of the outdoor cats it studied accounted for three-quarters of the creatures killed—meaning, as the researchers wrote, that "a large fraction of owners considerably *over*estimated their cat's predation." In fact only one cat was an ace bird killer; the rest made do with mice, voles, and "undetermined innards." The researchers also note, with calm neutrality—a quality, perhaps uniquely Swiss, present seldom if ever in the innumerable studies I've read on cat depredation—that "prey species on continental landmasses have co-evolved with domestic cats over hundreds of generations and have thus been considered little susceptible to this hunter." Nonetheless they do add, studiously, that "recent declines in many farmland and garden birds, the importance of gardens as wildlife refuge[s] in fragmented landscapes, and increasing cat populations due to intensified urbanisation have brought the ecological role of cats on continents into focus of much scientific debate."[8]

Since for the most part you or I don't know whether our particular cat is a hunter or just a looker, we must revert to the chilling core finding of the meta-study by Scott R. Loss and colleagues cited in chapter 5: somewhere between 1.4 and 3.7 billion birds (and 6.9 to 20.7 billion mammals) are killed by cats every year in the United States.[9] Loss et al. didn't break down how many were killed by pets and how many by ferals. But let's give our own cats a ridiculously huge benefit of the doubt and say the feral cats committed 90 percent of the crime. That still leaves us owners of outdoor cats with between *one hundred forty million* and *three hundred seventy million* dead birds on our hands.

Math can be funny, though. Let me try to lighten our shame. Stick with me through the arithmetic. Because all the self-respecting humane organizations—at least the American ones—are so strongly in favor of keeping all cats indoors, there are no reliable statistics about how many American pet cats are indoor-outdoor. Advocates of keeping cats indoors say 65 to 70 percent of American cats are free to go outside at least some of the time.[10] A more scientific but small-sample study reports that half or fewer are indoor-outdoor.[11] In other countries, far more cats are free to stroll around at least some of the time—more than 90 percent in the United Kingdom.[12]

Since so many American cats live in apartments and other places where there's really no choice about letting them out, let's say that the proportion of indoor-outdoor cats is half. If we take the generally accepted U.S. population of pet cats as ninety million, then forty-five million have some outdoor freedom. The Loss study's numbers, then, come out to 3.1 to 8.2 dead birds per indoor-outdoor cat per year. And that includes farm cats and all sorts of other unmanaged cats. If you are taking good care to minimize your cat's depredation—a subject we'll get to—you can cut those numbers in half. So take the median between 3.1 and 8.2—5.6—and halve it to 2.8. Two-point-eight birds per year! That's not even a quarter of a bird a month! Good kitty!

(What was it I was saying about finding what you want to find?)

In fact there is a lot you can do. Careful scheduling of liberty hours protects not only cats: At dusk, at dawn, at night, ground-feeding birds may be particularly vulnerable, and all the little four-legged critters are likely to be most active at those times as well, so that's when to keep the cat inside. Teaching your cat to come reliably when called—by reliably administering treats at each prompt return—can head off all manner of mischief.

You say you live on the fifty-seventh floor on 57th Street? There's Central Park virtually at your feet. Teach the little furball to walk on a leash! "It is best to start as early as possible, before your cat has developed a fear of the outdoors or a fear of unusual noises," say the experts at PetMD.com, insisting that it can be done. "Older cats are often more reluctant to go outside on a leash—or to be on a leash at all"—no kidding!—and "It may take months to get her used to accepting a harness, and to being led, but with diligence and a wish to succeed, you can do it."[13]

My former wife and I, just married at 20 and 22 years of age respectively, and residing on the top floor of a little brick house on West 95th Street in Manhattan, adopted a tuxedo-marked kitten in Greenwich Village and named her Elvis Abdul Ho Chi Minh McNamee the First (look, that was the tenor of the times). Our house was the only one on the block with a stairway to the roof, and soon Elvis was bounding from parapet to parapet flat out forty feet above the street, eyes alight with the joy of flying. After days of confinement, with both of us away at work, Elvis would grow cross, liable to lash out and scratch when approached, but a big bowl of Tabby Treat and a run on the roof would restore her constitutional sweetness. Louise wrote her a theme song, to the tune of "The Ballad of Davy Crockett":

> Born in a tenement on West Fourth Street
> Big long tail and four white feet
> Raised on T-Treat so she knows what to eat
> Mean and ornery but no one can beat
> Elvis! Elvis McNamee!
> Queen of the Upper West Side

Weekends, we went to a cabin in southeastern Connecticut. In the apartment, Elvis would hide at first sight of suitcase or swimsuit—she hated to ride in the car. But as we approached North Stonington she came alive, and she would spend the weekend prowling in the ferns, gobbling up lobster guts, cornering mice on the screen porch, snoozing by the fire. Wherever we went, her whole life long, she never wanted to go at first, because she knew that meant being put in the car and the misery of the ride (whyever that was—we never knew), but once we got there she was happy-happy. Then, at the end of the outing and facing another car ride, she didn't want to leave. We took her to the Catskills, where she had a fine time all weekend till it was time to go and she crammed herself into a heating duct. We took Elvis camping in Virginia: She rode down there like a champ, seven hours in the back of a jouncing Land Rover amid piles of sleeping bags and clanging pots and pans, and all weekend she was content just to patrol our little campsite, but as we struck our tents we barely caught her before she fled into the forest of Shenandoah National Park. We made the mistake of staying into the first day of bear season in Vermont, where the first gunshots she had ever heard drove Elvis to ground and it took us into early dark to find her, hunched and stone-still in dense cover. We took her to St. Croix, where she hunted lizards and marveled at the flocks of yellow banana birds. We took her to Memphis every Christmas, where she always stayed at Louise's parents' house (mine had dogs). Elvis loved Louise's childhood cat on first sight, another girl cat with a boy name, Napoleon, the placidest feline imaginable: Their great pastime was to lie looking at each other for hours—until one day, on who knew what whim, Elvis took to a pecan tree, fell out, and dislocated her hip, the only mishap she ever suffered in eighteen years of outdoor adventures.

This wasn't entirely dumb luck. We did keep close watch on her. We kept close watch on Augusta, too, but obviously not close enough. That summer of rattlesnakes, she ran away. Or disappeared. Was she out there dead somewhere? If she was dead downwind, she could have been less than a hundred yards away in the sagebrush and we wouldn't have known it.

Cats do run away. (Damn you, Augusta!) If you choose to grant your cat the freedom of fresh air and green earth, that is among the biggest risks you take. And it's not only your own risk, of course, the risk of losing your cat; it's a risk to the cat, too, of falling victim to any of the hazards of freedom.

Remember also, however, that indoor cats disappear too. (In a 2007 study of lost cats, 41 percent had been kept indoors.[14]) Who hasn't seen an indoor cat gazing longingly through impassable glass? And we all know that when a door opens, virtually every cat in the universe will by instinct want to go through it—even into a motel hallway, a Walmart parking lot, a cornfield, a forest. Why? Why do cats fall out of high windows? Don't kid yourself about them spread-eagling themselves into feline flying squirrels and soft-landing on the concrete from two hundred feet up. (There are such stories, nearly all apocryphal. The true ones are vanishingly rare.) What is true is that they can right themselves and land on their feet, but from two hundred feet, or even one hundred, unless there's an awning or a pond, that will be a bone-crushing fall and horrible.

Why do they run away and get lost? Of course, sometimes they come back, sometimes almost miraculously. Nearly a century ago, Francis H. Herrick conducted a series of experiments in which he took a number of cats several miles from his home in Cleveland, blindfolded, boxed, tied in an opaque cloth bag, and even, in a few cases, anesthetized by chloroform; and they unerringly found their way back, sometimes over very daunting obstacles.

You can find all sorts of stories of astounding cross-country homings, some of them probably true. Roger Tabor, a British biologist, has cited several that he resolutely believes: "Murka, a tortoiseshell in Russia, traveling about 325 miles home to Moscow from her owner's mother's house in Voronezh in 1989; Ninja, who returned to Farmington, Utah, in 1997, a year after her family moved from there to Mill Creek, Washington; and Howie, an indoor Persian cat in Australia who, in 1978, ran away from relatives his vacationing family left him with and eventually traveled one thousand miles to his family's home."[15] Much more common are the heartbreaking disappearances. Many of those, however, are avoidable.

First of all, let's hope you've taken care to have three critical things in place before the cat was gone. The first will probably mean that the cat will be back in short order: a collar with an identification tag that includes the cat's name, your phone number, and your email address. The other two are an implanted microchip (registered with a tracking company) and a good current photograph (to put up on posters and flyers and the internet).

If you suddenly look around and the cat is not to be seen, as soon as you catch your breath you can calm yourself with the knowledge that most cats don't go very far.[16] In fact, there's a decent chance the cat isn't gone at all.[17] Augusta several times found the most unbelievable places to hide and then would absolutely neither move nor make a sound. We had no idea why. She abhorred vacuum cleaners. At the ranch, she tore a hole in the bottom of the box spring in the guest room and would climb up into it and perch on one of the slats until the house cleaners left. When we finally found her, she never resisted, indeed seemed happy to be found. So: Really, really search your house, softly rattling a bag of treats as you go.

If he's not inside, the odds are he may still be nearby. If you see him, he'll probably be jumpy. Don't run after him, don't holler his name, don't clap your hands. If he looks at you, fall to your knees, look away, and maybe give him a brief shy peep. Maybe whisper his name. Chances are he's scared about something. Reach out a finger—he may come and sniff it, and then you're both home free.

If he keeps going, though, and really is gone, you need to set up some sort of welcome center at home, outside if possible—some old clothes of yours, the stinkier the better, inside an upside-down cardboard box with a hole cut in it just big enough for the cat to get in through. Put out food, water, and a litter box with used litter in it.

Plastering the neighborhood with posters and slipping flyers under every door can be quickly successful. (Don't put them in mailboxes. That's illegal.) Some people can find the strangest reasons for deciding that your "stray" cat was starving or lost or abandoned—so if the flyers and posters haven't guilt-tripped them into coming clean, ringing doorbells and looking them in the eye may be an effective next step. Put your posters on bulletin boards wherever you can find one—especially any local veterinarians' offices. Notify the municipal animal control agency, the SPCA, every volunteer animal organization you can think of, and the police. Facebook and Instagram pages can be invaluable.

Naturally, you want to go looking, too. Late at night—like two or three in the morning, or whenever traffic finally winds down—is the best time. If the cat makes the slightest sound, you'll be able to hear it. Carry a bag of treats to shake, and maybe a few cans of food as well—the sound of an opening can of cat food can be magic.

Put a lost-cat ad in the paper and on your online neighborhood bulletin board. Check the papers every day to see if somebody reports finding a cat. If you've moved lately, make sure the cat didn't go back to his old home. Sometimes you can borrow a Havahart trap from the SPCA, or you can rent or buy one, then bait it with your cat's favorite food and set it out in the yard. You may well catch a feral cat, somebody else's cat, a raccoon, a skunk, who knows what—but you might catch your cat, too. Try to think of other things you might do to make your home recognizable and attractive, especially odorants—old shoes on the front steps, say.

Keep trying. In the 2007 study by Lord et al., referenced previously, the owners of lost cats did a pathetic job of trying to get their allegedly beloved kitties back. Only 14 percent of the cats had any form of identification. The median time before any of the people got in touch with an animal agency was three days, and for those who went to an agency more than once, the median interval was eight days—so because the holding period at most of the agencies was three days, at least some of the lost cats were euthanized. In the end, barely more than half of the lost cats—53 percent—were recovered, and two-thirds of those simply came home on their own.[18] No thanks, in other words, to their human guardians.

The lesson here is that if you *care*, if you *try*, your chances of getting your lost cat back are really pretty good. Keep looking. Keep calling. Keep checking the agencies and the vets' office. Maintain your ad in the paper, and your social media posts. When somebody has torn down one of your posters, put up a new one. Talk to kids around the neighborhood. (Little girls are the best lost-cat finders of all.)

Never forget that your cat has very probably not gone far. Try not to broadcast how freaked out you are—in fact, see what you can do to calm down—because if your cat comes near, he can read every slightest quiver of your body language and the faintest whiff of your anxiety scent.

And even when it seems all hope is lost, remember this: Sometimes they stay away quite a while, and then they just…come home.

Case in point. A cat named Holly disappeared at a Good Sam recreational vehicle rally in Daytona Beach, Florida, in November 2012, two hundred miles from her owners' house in West Palm Beach. They searched for days, distributing flyers, checking the agencies, all to no avail. Almost two months later, on New Year's Eve, Holly was spotted in a back yard about a mile from her home, her feet worn to nubs, the pads bleeding. She was down to half her

normal weight, and barely able to stand. The woman who found
her fed her for a few days and then took her to a veterinarian. The
vet checked, routinely, to see if Holly was microchipped. She was.[19]

American humane associations are unanimous—fiercely so—in insisting that cats be kept indoors. The reason is single: safety. But there
are costs, too.

We have learned that the inner nature of *Felis silvestris catus*
is inextricable from the species' wild heritage. More recent evolution in the company of human companions has added to that
nature, but not erased it. When the cat's life comes indoors, with
it come a number of instinctive and ineradicable realities, in particular[20]:

- stalking, seizing, and attacking prey;
- patterns of play that emulate those;
- a need for privacy;
- curiosity, just as the proverb holds;
- a need to scratch, to climb, to observe from a safe refuge;
- territoriality, evinced in part in what may seem an almost
  obsessive concern with the location and hygiene of the litter box;
- a need for social relationships with other animals, including
  human ones and, if possible, other cats;
- a need for consistency, reliability, predictability, affection, and
  trustworthiness in the behavior of the cat's human companions;
- a world invisible and inaccessible to us, defined by scent, especially pheromones;
- the fact that a cat has a mind, which needs communication,
  stimulation, puzzlement, interest.

If our cat is to have a good life indoors, we must meet all those needs and accept all those facts.

The most wonderful addition to the heritage of our cat—"recent" in that it seems to have emerged through association with humans— is that the subspecies *catus,* unlike its desert ancestor *Felis silvestris lybica,* has evolved a heart designed for love. This unique animal thrives on love, and loves in return.

Indoor or outdoor, a cat can be deprived of love, or loved. Love is not essential. But it is, I must say, advisable. It will make life with your cat, oh, maybe a hundred times better than without it.

As for the critical needs listed above, many are the cats whose human companions don't know that those needs are *critical.* Obligatory. Theirs are the cats—we've all seen them—who are listless, lazy, overweight, in effect semi-animate sofa cushions. Prisoners of boredom. More item of décor than beloved companion.

The outcome of denying an indoor cat's nature is likely to be more than just ennui. It is what the cat whisperers thrive on. These are the cats who rip upholstery, marinate carpets, spray the walls, yowl at five in the morning, scratch, bite, defecate on your pillow—and, sometimes, given the chance, run away.

Yes, there's a lot for the conscientious owner of an indoor cat to do. You really should play with your cat every day, but it can become routine and fun. And the physical configuration of your cat's habitat, as discussed in chapter 3, can make a great and happy difference. If you live in the Eighth Arrondissement amid walnut boiserie, crystal sconces, and Louis XV écritoires, you may not want Jackson Galaxy's plywood perches covered with shag carpet remnants at various heights along the walls, or his cat superhighway running all the way around the room just below the ceiling. Even a just sort of *nice* house is not esthetically improved by one of those ghastly cat condos you get at Petco or wherever, with their multiple levels of cheap-rug-covered platforms and little hidey-houses and dangling balls and such—but I gotta tell ya, cats love them. If you can stand to put one of those eyesores next to a window with a bird feeder outside, if you can find the sweet spot for the litter box and you clean it religiously every day, if you've got scratching posts modestly tucked away but actually all over the place, if the cat can find a cozy lookout post atop the kitchen cabinets or the refrigerator, if she's found her private refuge under a bed on that crumpled duffel bag that you keep thinking you were going to throw out…see? Things can start to fall into place without a lot of hammer, saw, and catification. The point is that even in your palatial Paris apartment, ways can be found to satisfy your cat's wild nature. You just have to pay attention to the principles.

These are all easy positives, but it's important to think of some positives also as absences of negatives. For most cats, fear is the biggest negative. They evolved not only as predators, remember, but also as prey, and tuned as they are to what may seem to them signals of possible threat, many cats are highly sensitive to distress—including some kinds of distress that may not occur naturally to us. Loud noises, for example, can be very upsetting, as can the mere sight of unknown cats outside. Little kids running straight at a cat can terrify him to the point of dangerous self-defense. The point of a high perch is not just so Mr. Kitty can look the proud lord but also so he can feel safe—it's part of his biological heritage.

Teaching your cat early to regard his carrying case as a refuge has the additional benefit of making travel much less stressful, even trips to the vet. Any room in your house can be made a safe space—especially if there are dogs in the house, or other cats—by installing an electronic cat door activated by a radio collar.[21] When there are more than one cat in a household, the possible stressors are many. Mieshelle Nagelschneider's book *The Cat Whisperer* provides excellent guidance for reducing inter-cat conflict.

Play is a complicated business, but the cardinal rule is that it should resemble the full hunting sequence as much as possible, including the kill. Endless pursuit is endlessly frustrating. Laser toys excite the chase instinct very effectively, but they don't provide anything to pounce on, and too many people wielding lasers innocently drive their cats to baffled, game-quitting apathy. A toy to grab and bite, or even just a little crunchy treat after a minute or two of laser chase, will satisfy your cat's craving for prey. I'm not sure what I think about cats watching cat TV or birds through a window for hours at a stretch, with no end to the game. I suppose it's better than boredom. They do seem fascinated, and they show no ill effects. A catnip mouse or a toy with treats inside that the cat has to struggle to obtain seems more to the point, though since there's a reward in store instead of incessant frustration. There are some excellent interactive toys—some as simple as a ball in which you hide crunchy treats, requiring the cat to chase and bat the ball around to get the treats out. In all cases it's crucial to remember that play is likely to be the only exercise that most indoor cats get.

Playing with *you* is much more important than what the game is. It's an act of love for both of you, and it takes no equipment more than a string, a ribbon, a paper bag, a cardboard box. Cats will chase, fly, dive, scurry, hide, pounce from hiding, wrassle, twist, jump, flip, fall, flee, race, attack, scare themselves for the fun of it, end up panting flat on the floor exhausted and so, so happy—and they will know you've done it for *them*.

Food, ah, food. Cats learn nothing more quickly nor thoroughly than using food to torment and control their human guardians. In its first line on Machiavellianism, Wikipedia takes us to the Oxford English Dictionary, which defines it as "the employment of cunning and duplicity in statecraft or in general conduct." Statecraft being beyond the reach of cats, the feline Machiavelli—which is all of them—concentrates on general conduct, with an emphasis on gastronomy.

Up to a certain point, we do have to meet their demands. The species is an obligate carnivore. That means they have to eat meat. Their digestive tract is very short, so vegetable matter passes straight through, undigested. They have high requirements for vitamin D, niacin, and vitamin A because they lack the enzymes that other animals' digestive systems provide. Their feline ancestors, for hundreds of thousands of years, ate no plants at all. They subsisted entirely on a diet of the animals they killed, and the domestic cat is anatomically and metabolically identical to them.[22] It doesn't matter if you're any kind of vegetarian, your cat cannot be one. You can't use meat substitutes. Dogs, actually, can live on a vegetarian regime. Cats can't. They'll die.

That does not mean, however, that way out on the other end of the spectrum, the people who rip up raw chicken and goat meat and suchlike for their cats are doing the right thing either. Wild cats—and ferals—eat their prey whole: fur, feathers, guts, teeth, skulls, stomach contents, the works. With those come vitamins, minerals, and all manner of micronutrients that the folks who think they're doing their so-wild beasties such favors are depriving them of. The beneficiaries of this ill-informed kindness are pretty well guaranteed to end up quite sick.[23]

At the specialty shops there are cat foods in which the first-listed ingredients—those of highest volume in the can—are in fact flesh, and they cost a lot of money. Maybe they're better, I don't know. We started our dear Isabel out on those things, and she was delighted, and healthy, but her sustenance was getting toward more expensive than our own. We asked our vet what he fed his own cats, and he said Friskies. The names of the varieties of Friskies (as of all other plebeian commercial cat foods) are perfections of legalese dodge-'em: "With Beef in Gravy," "Mariner's Catch," "With Salmon & Beef in Sauce," "Sea Captain's Choice," "SauceSations Turkey & Giblets Dinner in Homestyle Sauce." But Purina, the parent company, promises on the label that every one of its canned cat foods is "formulated to meet the nutritional standards established by the Association of American Feed Control Officials (AAFCO) Cat Food Nutrient Profiles for growth of kittens and maintenance of adult cats." I immediately suspected that AAFCO was a front, made up of feed company representatives. In fact it's a voluntary coalition of government agencies, with nothing to lose or gain from their rulings.

Then there's the dry stuff, another matter entirely. Both wet and dry are considered nutritionally complete, but obviously with dry food you need to be sure your cat is getting enough water, especially because kidney disease is one of the most common serious illnesses of cats. Some people mix wet and dry. Some cats are staunchly opposed to dry food, but it is convenient, and although the wet-food crowd may look down their noses at it, it's perfectly adequate. Or at least that's the latest information—a few years ago the story was different. Talk to your vet. In any case, what is most important is to maintain a diet without too many surprises. Cats like consistency and predictability in all things.

This is where the Machiavellian manipulation can start to come in. Cats may love one food one day and hate it the next. Harried owners with soft hearts can be seen (or will hide while) opening can after can, as His Highness sniffs and walks stiffly away from each in turn.

It may not be so simple after all, however. Little Machiavelli may have a case. There's new research showing that cats' olfactory systems are exquisitely fine-tuned to the nutritional makeup of their gustatory choices. The researchers started by offering their feline subjects several sorts of food, and the cats made their initial choices based on the flavors they preferred. Fish and rabbit were favorites. But over time they began to select foods with a particular protein-to-fat ratio, and they would even choke down *orange*-flavored food in preference to the "good" stuff if it had the right energy balance. The cats chose 70 percent protein to 30 percent fat across the board, regardless of fragrance or flavor.[24]

The question that now arises is whether the Prince can detect micronutrients as well, and in refusing both the "With Beef" and the "Sea Captain's Choice," or whatever, is making informed dietary decisions. Maybe future research will reveal that cats' food fussiness is genuinely about nutrition. A big maybe, but if so, it will represent a major victory for picky cats everywhere, and will hit cat owners hard in the wallet.

The major point about cats' food in the meantime—for us—is that it is the foundation of discipline. For cats, it is the great opportunity to learn that you can be manipulated, and yes, they are capable of generalizing the principle. We must never forget that as sweet and nice as they can be, cats have no conscience whatever. No sense of justice. Fairness, to a cat, is not a recognizable concept. This does not mean they are bad, or cruel. They do take pleasure in torturing their prey, it's true, and nobody seems to know why. But we are not their prey. We are their cohabitants, and with (as we have seen) assiduous and gentle effort we can be their friends, and teach them to be ours.

They never forget, however, that we control the food. They may learn to open the cabinets, and they can certainly learn to rip open bags of kibble, but they can't open cans. They can't go to the store and buy food. This is our true power, and we can use it for good. You may choose to feed on a regular schedule, or "free feed," with food out all the time as long as the amount is carefully regulated. Whatever you do, all you have to do is be consistent and the cat will get it. This is kindness, not cruelty.

Unlike our cats, we are eminently capable of cruelty, and we can hold anger in our hearts for years. Some people who claim to love their cats also hit them. Kick them, throw them across the room. Yell at them and scare them. Cats are very easily frightened, and one of the few emotions they will hold in their hearts for years, as we hold anger, is fear. Don't scare your cat, *please.* You control the food, that's all you need. I'm not saying withhold it to show your power. I am saying that once you let your cat beg and wheedle and use false affection to inveigle you into just a little more, Papa, please? then the old-time word is the right one: spoil.

There is also the profound justification of health. Just as is the case with contemporary humanity, excessive body weight disposes cats to a wide spectrum of ailments, from diabetes to arthritis to heart disease. Fat cats, like fat people, tend to die early. This isn't hypothetical, either. Fifty-eight percent of American cats are overweight or obese.[25] The percentage for indoor cats alone (versus outdoor and indoor-outdoor) hasn't been isolated, but you can safely bet it's higher than that.

Discipline becomes its own reward. When you start to use those little crunchy treats to reward your cat for doing what you wish her to do, and you use the clicker to precede that reward—with the click instantaneous, so that the cat cannot fail to understand what it's about—you will feel your power and so will your cat, and both of you will soon come to reap real satisfaction from the reliable, symmetrical dance of action, click, and treat. It's not in the least complicated, but if you're willing to keep at it, you can teach your cat to do extraordinary things.

You may not want to turn your kitty into a circus performer, but clicker training is how it's done. The doyenne of the clicker is Karen Pryor. In her book *Getting Started: Clicker Training for Cats* and at her website www.clickertraining.com you can learn that with persistence, patience, and consistency, you can teach your cat to do almost anything a cat is capable of, with a spoken command or a hand gesture—the feat to be followed instantly by a click and then a treat.[26] That simple. Pryor also clicker-trains dogs, horses, birds, and—for real—rabbits.

There's now an unbelievable sort of cat sport called Feline Agility Training, in which cats jump hurdles, shoot through tunnels, bound up stairs, and weave around poles all within a thirty-foot-square course, racing against one another for doing it fastest—and they don't even get clicked and treated. The only way they're trained is by chasing a fluttering bird toy or some such. The champs can blast through the whole obstacle course in under fifteen seconds. Like I say, it's unbelievable. But it's real. Take a look-see at agility.cfa.org/index.shtml. And you thought Fluffy could never learn anything.

Perhaps the best thing you and your cat will ever learn together is the word "No." Click-and-treat, again, is the secret here. A sharp *No!* and if the cat even pauses, you give her a click and then a treat. The technique is a matter, entirely, of successive approximation. Click by click and treat by treat, the pause becomes a hesitation, the hesitation becomes attention, attention becomes full stop. Once she has wholly absorbed it, once she feels it in her being, you can save her life. A panicked cat will do idiotic, insane things—jump out a window, plunge into traffic. Teaching a dog to stop is like teaching dogs lots of things, not so hard, and you don't get a trophy. Once you've taught your cat to stop, you deserve a reward. Maybe this is reward enough: You get to keep the cat.

From discipline arises attunement. Accepting your authority, your cat comes to trust you. Limits and boundaries become definitions of safety, and a cat can come to understand that fact. Once that understanding is established, you will find that when fear strikes—thunderstorm, woofy dog, grabby child—he runs to you. Food may always be the primal urge, deep down, but once it becomes a settled routine, and once trust is in place, what comes to the surface is the physical expression of calm and affection, most expressive in the form of touch. Cat head-bumps you, rubs chin on your chin, wraps tail on your shin, lies with cheek to your thigh. Gives you the long, slow blink, the quick flick of the tongue.

I've said this already, but it's worth saying again: *Cat teaches you.* Which stroke where, which direction, how hard, when finger when palm, yes there not there, okay, enough. You can get better at this for years if you keep paying attention. Pick me up, walk me around, I like looking out the window in your arms much better than sitting on the sill—look at that bird! (Feel his skin twitching all over?) *Paying attention.*

There are lots of books about relating to companion animals, but there's at least one that I guarantee you your cat will thank you for reading at least the cat-relevant parts of: It's called *The Tellington TTouch* (why the double T? I'm not sure), by Linda Tellington-Jones.[27] The TTouch is actually easier to learn visually than from the book, and fortunately there are both a DVD[28] and a number of YouTube videos. It's basically a massage technique. The author started it with horses, and in the book there's a quantity of New Age mysterioso to wade through, but the thing works like magic on cats. It does get pretty technical:

> Alternate the Clouded Leopard and Lying Leopard TTouches beginning at the base of the neck. Move randomly from place to place on your cat's body and make each circle fairly slowly....Begin with a number three (see page 26) and experiment with a firmer or lighter TTouch.

Don't worry if it sounds bogus, I'm telling you it's worth a try.

The American Association of Feline Practitioners and the International Society of Feline Medicine have made an attempt to define what is a good life for a cat by boiling down the essentials to what they call The Five Pillars of a Healthy Feline Environment. They recapitulate some of what's already been said here, and may not be quite complete, but the simplicity and clarity of the Five Pillars go to the heart of what a cat cannot do without:

1. A safe place.
2. Multiple and separated key environmental resources: food, water, toileting areas, scratching areas, play areas, and resting or sleeping areas.
3. Opportunity for play and predatory behavior.
4. Positive, consistent, and predictable human–cat social interaction.
5. An environment that respects the importance of the cat's sense of smell.[29]

"Positive social interaction," as a term, may have the unfortunate smell of the clinic, but readers of this book will have understood by now that a much deeper emotional connection than that phrase implies is as indispensable as an adequate physical environment. And a dull life, a life of boredom and ennui, is not a good life. A cat's life cannot be fully good without mental and sensory stimulation, and it cannot be good without love.

༝

Some people, contemplating so much responsibility, may well think there ought to be a better way. How about breeding a better cat—one without so many and such exacting requirements? Stephen Budiansky, in his book *The Character of Cats,* recalls the interesting fox-breeding experiment by the Russian biologist D. K. Belyaev, who "demonstrated that by selecting for nothing but tameness, such as a cub's willingness to be approached and handled by a human being, it was possible in just five generations to produce a strain of foxes that had acquired the whole domestication package. These foxes had piebald coats and drooping ears, they wagged their tails and barked just like pet dogs, and they whined and begged for attention just like pet dogs."[30]

In his book *Cat Sense*, John Bradshaw proposes the selective breeding of an "ideal indoor cat...which might be better suited to indoor life, or be more sociable, than any of today's cats." After discussing a range of impractical alternatives such as "hybridization with some of the smaller South American [wild] cats" (a nasty road this narrative has already been down), he settles on existing variation within the domestic cat species as "the best starting point for the completion of the cat's domestication." The technique "should be to identify those individual cats with the best temperaments, and to ensure that their progeny are available to become tomorrow's pets...."

"Selection for the right temperament among house cats requires deliberate intervention," writes Bradshaw; "natural evolutionary processes, which have served the cat well so far, will not be enough." What we're doing now, "neutering cats before they breed...likely favors unfriendly cats over friendly....When almost every pet house cat has been neutered...we must fear for the next generations of cats. These will then mainly be the offspring of those that live on the fringes of human society...." The ultimate result, then, of depending mainly on the adoption of shelter cats, as we do now, with so many of them drawn from feral populations, will be, in Bradshaw's view, "pushing the domestic cat's genetics back gradually toward the wild, away from their current domesticated state."[31]

Budiansky, contrariwise, maintains that if cats were going to be truly domesticated—predictable, obedient, non-predatory—surely, at some time over the last ten thousand years, they would have been. At least a little.

I don't know how you feel about this idea of eugenic improvement of cats. Isn't it an essential part of catness to be sort of a pain in the neck?

✣

There are terribly many ways to misunderstand cats, equally many cats' lives that are not good. Almost always it comes down to denial of their nature—as if they didn't even have an inner life. The philosopher Bernard Rollin makes the fundamental case. Besides all the good that having pets does for us, "we must also remember that they possess intrinsic value, value that does not derive from their utility for us but is part and parcel of their moral status as living, feeling, sentient creatures; creatures whose lives matter to *them*." Once we grant them that status—the value of their existence—then we have engaged in a contract. By protecting, feeding, sheltering, and loving them—owning them—we owe them their lives. We owe them good lives.

Rollin has no mercy for the many who "violate this contract in the most essential ways"—the people who return cats to shelters knowing that except in the relatively rare cases of no-kill organizations like the San Francisco SPCA, there's a good chance (41 percent[32]) that the outcome will be euthanasia; the people who bring their cats to veterinarians to be put to sleep (100 percent) when there's nothing the matter with them:

> People bring animals in to be killed because they are moving
> and do not want the trouble of traveling with a pet. People
> kill animals because they are moving to a place where it will

be difficult to keep an animal. People kill animals because they are going on vacation and do not want to pay for boarding and, anyway, they can always get another one. People kill animals because their son or daughter is going away to college and cannot take care of it. People kill animals, rather than attempt to place them in other homes, because "the animal could not bear to live without me." People kill animals because they cannot housebreak them, or train them not to jump up on the furniture, or not to chew on it, or not to bark. People kill animals because they have moved or redecorated and the animals no longer match the color scheme.... People kill animals because they feel themselves getting old and are afraid of dying before the animal. People kill animals because the semester is over and Mom and Dad would not appreciate a new dog. People kill animals because they only wanted their children to witness the "miracle of birth," and they have no use for the puppies or kittens.... People kill animals because they are tired of them or because they want a new one. People kill animals because they are no longer puppies and kittens and are no longer cute or are too big....

Equally intolerable from a moral point of view are our flagrant violations of the pet animals' right to live their lives in accordance with their natures—natures we have shaped.

Rollin is on such a tear that once having admitted that ignorance is the main reason people don't grant their pets "lives in accordance with their natures," he's accusing them of stupidity as well, and culpable stupidity at that: "The average person who acquires a dog or cat is worse than ignorant, worse because such people are invariably infused with outrageously false information." You kind of want to say, Hey, professor, how about a walk outside and a nice cup of tea before we go on with this?

Fact is, once cooled, he's right. Hence this book.

Some of the most egregious violations of catness come in the ever-proliferating form of cat videos. Let me quickly say that they're not all bad. But even the cute and relatively harmless ones often dwell on deformity or fear. The highly respected Walker Art Center of Minneapolis founded the International Cat Video Festival in 2012 and curated it with care, in order to exclude the sadistic and highly manipulative exhibitions that characterize so many cat videos, and drew more than ten thousand people to the first festival. By the next year the festival had become a tour of fifteen cities, and its stars were internet sensations like Grumpy Cat, Lil Bub, Pudge the Cat, and Old Long Johnson.[33] Grumpy Cat's ha-ha permanent frown is due to an underbite and a birth defect. Old Long Johnson seems to speak in human-like sounds, but it turns out that there's an off-camera cat annoying and scaring him; he is prevented from getting away, and those are his particular expressions of anger and fear.[34] Many feline internet stars are physically deformed in some way—often harmlessly, it's true—and their admirers' reaction to them is sympathetic at one level and mocking at another. It's a strange brew. But it's a very successful one. No one seems to have any definitive measurements of its success, which strikes me as extremely odd, considering that YouTube is owned by Google, the universal master of big data. YouTube does say that it alone has more than two million cat videos at its command. As of October 2014, the last date for which there is published data, they had attracted some 24.6 *billion* views.[35]

Cats dressed up as Santa Claus. Cats falling off porches. Cats with people waving their front legs so they can "conduct" some music. Cats running into boxes, as we all know they will do, and getting stuck, and nobody helping them out but rather making a video of it. Cats seeing themselves in a mirror, jumping to attack "the other cat" and smashing their heads against the glass. Cats confused. Cats angry. Cats scared shitless. Ha ha. And I know, sometimes the video is perfectly innocent, just funny, but I'm sorry, the whole thing is beyond me, and beyond the scope of this book, except insofar as you, the reader, may well be one of those viewers, and I, the writer, am here with a gentle caution to think about the experience of the cat in the video.

Every cat has a right to a life in accordance with her nature. Look at the video and ask yourself, Is this cat's life a good life?

<p>

Thirteen days after she disappeared, at six o'clock in the morning, Augusta appeared on the bedroom windowsill, meowing to be let in. I was still back East, in New York again. Elizabeth called, in tears, barely able to speak. She had burst outside and run to the cabin window in her nightgown, barefoot, and grabbed Augusta. She kept clinging to her, hard, as she rushed back in and searched for Augusta's food and her bowl, which in despair she had already put away in a cabinet.

The next day, a deer jumped over a pasture fence and half across the hood of Elizabeth's car as she drove home from a visit with friends. The front end was crushed and the deer was dead. There seemed to be something missing in our conversation. Why had she been away instead of home with Augusta? Everything seemed, somehow, still awry. I didn't come home right away, I waited three days. Coward, fool.

Augusta was fine. Skinny. Fur funky. But ready to resume her routines, and sweet as ever. What did you do, little doodoo head? Where did you *go*?

Across the river, across the bridge that Augusta had been crossing that night when she may have thought we had abandoned her, there was a house with a part-time caretaker who had been mowing the lawn on the day Augusta disappeared. He kept the lawnmower in a shed and the door was open. The next day he was gone and the shed door was closed, locked. We had called him—he lived in Billings, as did the owner—and he said, no, he was sure there had been no cat in the shed. I remember now he was kind of a jerk, irritated by the question. Why hadn't we somehow found a way to get into that shed and check anyway? At least call through the door and then listen? It was too near the noisy river—we wouldn't have heard her soft voice. So I told myself.

The day Augusta returned, the caretaker also had returned. We believe, now, that what happened was that he had unknowingly locked her in that shed and she lived on mice for almost two weeks. Augusta had not meant to run away. So I tell myself.

Chapter Seven

Illness, Aging, and Death

Our cats' wild ancestors evolved with a talent for concealing illness, injury, and pain. To reveal weakness would be to invite attack. Despite having virtually no predators in their modern history to contend with, our cats have retained that talent. One might argue that it's no longer useful, that in fact it is distinctly disadvantageous, since the reproductive capability of the species would surely be enhanced by the aid and comfort that modern medical care could provide if only the cats would let their compassionate human companions know when they were in need; but the era of human caring has been hardly a minute of evolutionary time. This is a perfect example of the survival of our cats' wild ancestry inside the thin shell of what we like to think of as their domestication.

If Augusta showed any sign of her hip dysplasia in the summer of 2004, we missed it. We had rented a cabin on the East Fork of the Boulder for the second year in a row even though we didn't like it, but it was all we could find. We ached with nostalgia for the shady old-growth cottonwood forest on the Main Boulder that Augusta loved so much, where she spent so much time every day we had come to call it her office. The East Boulder place was all sun and dust, and Lord, it was hot that August. Yet Augusta spent long days on the roof, in the sun—"frying her poor pea brain," we laughed. Dumb kitty. Her black head, when she came down, seemed almost too hot to touch. She had good reason to perch up high: Once again, as in that bad summer on the lower West Boulder, there were rattlesnakes around, some of them big ones. There was one time, mid-month, when she didn't come home in the late afternoon, and dread rose in my throat like a blade of ice. I looked everywhere. Dust, heat, sagebrush, all I could think of was rattlesnakes. Friends had asked us to dinner—forty miles north, typical Montana—but now of course we had to cancel. Darkness came on, and I continued with a flashlight, calling, calling into silence. Surely, no, she couldn't have disappeared again, surely no.

At last, about to give up for the night, I shone the light into a corrugated steel culvert under a berm, no bigger than eight inches in diameter. There was the green-gold eyeshine of Augusta. She had drawn up into the smallest ball she could make of herself, equidistant from each end, and no, she was not coming out. She was terrified. I hypothesized that she had found refuge there from a coyote or a pack of them—the pipe was too small for a coyote, and cool, too—smart kitty! But please, kitty, come, it's dark and I'm getting cold. And finally she did.

From then on, Augusta made no further excursions, but she continued to climb to the roof to fry her brains. There were plenty of perfectly safe places where she could have perched in the shade. Could the warmth have been relieving pain in her hip? It didn't occur to me then, but now I wonder.

Cats seem to get old at different rates, and in general it seems that veterinary medicine is ready for whatever arrives—a relatively predictable bunch of problems, including not only Augusta's apparent osteoarthritis but also kidney problems of various kinds and the usual deteriorations that we all face. Little attention has been paid until recently to *successful* aging, to those wondrous cats who stay sleek and athletic well into their teens, sometimes even their twenties. The one thing that has been identified as an almost guaranteed no-entry card is obesity. Now at last there has been a comprehensive report by a panel of scientists on the whole complex of physical factors that go into becoming a healthy old cat. More than 20 percent of American cats are now older than ten, so this information comes at a good time. The report is highly technical, intended for veterinarians, and let's hope that it gets wide circulation among them. It lays out specific guidelines for assessing cognitive ability, medical conditions, musculosketal health, the state of the senses, dental and gum condition, gastrointestinal, respiratory, cardiac, renal, and endocrine systems; and a rich array of blood chemistry markers. It's called "Evaluating Aging in Cats: How to Determine What Is Healthy and What Is Disease," and I can't recommend it as light reading, but it is a major contribution to the collective life and happiness of cats.[1]

By that December, in San Francisco, barely nine years old, Augusta no longer ran like a rabbit downstairs for her breakfast, but rather hobbled—we called it boodling. Going up, sometimes she would pour on the speed. At other times, she would limp upstairs, and I could almost feel the grinding of her hip joint myself. Pain? Undoubtedly. Courage? No, just being herself. Which meant no complaints, just do it. Her spirit never flagged.

We started her on a nonsteroidal anti-inflammatory medication called Metacam, which seemed to help. Within a few months, nonetheless, she couldn't jump up on our high four-poster bed, so we found a rather handsome three-step mahogany library stair, which she took to with becoming dignity. It may well have been that being stuffed into her carrying case for a trip to the vet or the nightmarishly tiny box we had to squash her into for an in-cabin airplane trip to Montana caused Augusta agony, but she never once cried out.

They all hide their suffering. Step on Fluffy's tail, and sure

185

she'll screech, but if cancer is consuming her bone marrow and it hurts all day and every night you will probably not know it until her legs begin to fail and it's too late.

Still, it helps to pay attention. This book makes no claim to medical advice—there is plenty of that elsewhere—but it is worth remembering that cats are subject to a wide range of maladies that are difficult to detect and that can kill them with terrifying rapidity. The indications may be subtle. Hiding, loss of appetite, or loss of interest in playing or affection could be a sign of incipient illness. Recently an international panel of nineteen veterinarians with a wide variety of specialties identified twenty-five signs of pain—noting that all "were considered sufficient to indicate pain, but no single sign was considered necessary for it." The researchers caution that "The severity or intensity of pain…is difficult to define or quantify." The indicators, therefore, constitute only "a reasonable starting point."[2] They're still worth keeping in mind. Paraphrased to eliminate technical jargon, they are:

- lameness
- difficulty jumping
- abnormal gait
- reluctance to move
- reaction to palpation (applying light pressure with the hands)
- withdrawal or hiding
- absence of grooming
- playing less
- appetite decrease
- overall activity decrease
- less rubbing on people
- general mood change
- temperament change

- hunched-up posture
- shifting of weight
- licking a particular body region
- lower head posture
- squinting
- change in feeding behavior
- avoiding bright areas
- growling
- groaning
- eyes closed
- straining to urinate
- tail flicking

The veterinarian Jennifer Coates adds: "While this list is helpful, it only goes so far. For instance, a cat who has an abnormal gait might certainly be in pain, but other non-painful conditions (e.g., neurologic disorders) could also be involved. In cases where I have failed to find another reason for a cat's change in behavior and I'm left with undiagnosed pain as the most likely cause, I often rely on a tried-and-true veterinary test: response to treatment. I'll put my patient on a few days of buprenorphine—my favorite kitty pain reliever—and if his behavior returns to normal, we now know that pain is to blame."[3]

Researchers are finding other, new ways to assess pain in cats. Based on facial expression scales developed for assessing pain in babies, models of cat faces have been developed that use measurements of almost imperceptible changes in nose and cheek flattening and ear and whisker movement to assign a numerical score to pain intensity. Another approach, the Composite Measure Pain Scale—Feline, assigns numerical scores to behavioral changes, including vocalization, posture, attention to a wound, response to touch, and response to people.[4]

Something we can all watch for at home: If a cat urinates or defecates outside the litter box, watch her the next time she approaches it and see if she's having trouble getting over the side—that could mean pain in her limbs.[5]

Many problems, if caught early, can be quite effectively dealt with by a veterinarian. Taking your cat to the vet is not a hypochondriacal affectation. Even in the absence of medical symptoms, you should do it at least once a year. If you can find a veterinary practice that is cats-only, that's all to the good. For one thing, when you take your cat there—who may be freaked out to start with—she won't be surrounded by barking dogs in the waiting area. Perhaps more important, the cat doctor will have more experience than a generalist in diagnosing feline illnesses, injuries, and idiosyncrasies.

You also need to know where the nearest twenty-four-hour trauma center is. That will almost never be just a vet's office but a full-on hospital. Here's a quick list of things that call for *immediate* medical attention, from the newsletter of the Cummings School of Veterinary Medicine at Tufts University:

- repeated trips to the litter box
- labored breathing
- persistent vomiting
- seizures
- staggering or stumbling
- bleeding
- a symptom that persists or gets worse[6]

The international panel of veterinarians singled out three of their pain indicators that demand that you drop everything you're doing and get your cat to the vet right away: panting, pupil dilation, and squinting.[7]

A team at Tufts is working with other veterinarians around the United States—all with board certification in emergency medicine—to establish a network of what they call Level I and Level II veterinary trauma centers, so that vets anywhere can know quickly where to send a case in urgent need of a particular type of attention. Online access will enable rural veterinarians—who are usually distant from highly skilled trauma care—to connect to immediate information. Over time, thus, the country vets will learn emergency practice for themselves. The group is also pooling data on treatment outcomes from around the world—"a five- to ten-year vision for learning what works," says team member Kelly Hall of the University of Minnesota.[8]

The house we had found to rent in Sweet Grass County, Montana, in 2006 was Elysium for Augusta—far from the county road, with the swift Sweet Grass Creek running by to sniff along, a cottonwood forest full of downed and half-downed trees to prowl among for voles, rich green meadows crawling with mice. I was especially happy that the elevation was too high for rattlesnakes. There were coyotes out on the prairie, but as long as we made sure Augusta was well in before dusk and never out until full day, we were certain enough that she was safe from them: Their natural prey of rodents and rabbits was plentiful, and surely they knew through generations of life in ranch country that going near houses could be fatal. Coyotes are smart, and adaptable. A friend of mine in Sweet Grass County raises a band of a thousand sheep—with lambs—in the midst of whom a coyote family lives in peace, a détente with the big guard dogs having been reached long ago.

Within a couple of days of her arrival, Augusta's fur gained a sleekness it never wore in the city. On the prowl she traveled low like a leopard, eyes scanning, ears up forward then swiveling toward the slightest noise of possible interest or threat. She could jump the full three feet to the top of the rail fence and dance along it with no slightest bobble or limp. She brought us gifts living, dead, and in between.

Sometimes Augusta came in from her hunting with leaf duff, straw, spider webs stuck in her whiskers and fur. All you had to do was show the brush and say, "Brushing?" and she would assume the position—sphinx, little front feet together, head high, facing away from you, ready. She loved to be brushed, especially when, as it still was that summer, her coat was shiny and smooth.

She moved more slowly than in her younger years, and slept more. In a way, she seemed to love her sleep more—she appeared to relax more fully, and when she rose and stretched, although her arthritis may have been creaking a little, what her body and face expressed was contentment. The veterinarian Kathy Blumenstock considers the age of seven to be the frontier of seniorhood; Augusta would turn eleven in August. "If any species can elegantly accept the arrival of golden years," writes Blumenstock, "it's cats."[9]

When I was there alone, Augusta slept on Elizabeth's side of the bed, but still always, respectfully, at the foot. When Elizabeth was there, especially on Montana's chilly nights, she would stealthily find her way up to Elizabeth's knees, which would part just enough to accommodate a small, stretched-out sleeping cat. When Elizabeth turned to sleep on her side, she did so slowly, so that Augusta could ride with the turn, and maintained the cat-width gap between her legs, all this without waking. Augusta never considered attempting even the first part of this maneuver with me.

She nipped Elizabeth's bare ankles in the mornings, or, if she was wearing her backless slippers, her heels. She would follow Elizabeth in her robe with her tail straight up and her head cocked sideways and her mouth partway open. There was a particular look on her face. We called it bitey. Uh oh, she's looking *bitey*. Sometimes Elizabeth would drop a newspaper on the floor in front of her, *wham!* And then, when the moment of startle had abated, Augusta would settle for biting the hem of her robe. Sometimes, when Elizabeth bent over her, Augusta would bite her hair. She knew not to try any of that crap with me.

Back in San Francisco, Augusta began not to like her food. We would find a new brand, which would work for a while. Cantaloupe, when aromatically ripe, was always a success. I've never known another cat who loved it so—and cantaloupe only, not honeydews, crenshaws, Persians, not even the fragrant Charentais. Sometimes I cooked her chicken livers. She liked the milk left when you had finished your cereal. Elizabeth would take it to Augusta's feeding place in the kitchen repeating, *"Milk?"* in a tight little high-pitched voice, and she swore that Augusta's answering *Miew?* was the same word. Sometimes Augusta, *Machiavellissima*, would sneak to Elizabeth's cereal milk while it was still on the dining room table and get away with that. Her infirmities were buying her impunities.

Pets Unlimited is a big hospital, and sometimes, unless you want to wait forever, you see whichever vet is available. One veterinarian had prescribed Augusta's Metacam for her joint pain, and then in 2005 another added Cosequin, which slows the degeneration of cartilage. Every time Augusta went in for a visit—several times a year at this point—she would see a different doctor. We felt very fortunate in the summer of 2006, when Dr. Randy Bowman came up in the rotation. He made sure we understood his explanations, and he handled Augusta with extraordinary tenderness. We decided that from then on, whenever possible, Randy was going to be Augusta's doctor. Augusta's weight was down from her customary ten pounds to 9.6—nothing to worry about, Randy said.

That December her weight was down again, to 9.1 pounds. Her agility was going downhill as well. When Randy manipulated

Augusta's back legs, she cried out in pain. He decided to step up her medication to the steroid prednisolone, which would reduce the inflammation and, thereby, the pain. It would probably increase her appetite a bit, which would be good, but we should watch and not let her get fat, he said. It was also an immune system suppressant, so we should be on the lookout for infections. Diarrhea could be an effect, but it would probably go away. If not, we should call him. There were a lot of other possible side effects on the information sheet Randy gave us, some of them harrowing. As it happened, Augusta suffered no ill effects whatever from the prednisolone. She improved quite a bit, in fact. You could tell her pain had been greatly relieved, and her spirit remained indomitable.

Yet there began to be mysterious midnight visitations of—we didn't know what it was—some kind of terror? Augusta had begun to sleep sometimes in my office, in the high-walled circular bed behind my desk that we called the Bucket, instead of her customary place at the foot of the bed, and now sometimes, in the dead middle of the night, she would come out into the hallway, well down from the bedroom, and cry out loudly, the deep-pitched wail that she produced only when she was disoriented and lost in Montana. There, it meant, "Come find me," or at least, "Tell me where I am," and a good holler to orient her to the house or to oneself would usually be sufficient to bring her bounding home. But what was this? I would come to her and try to assure her that everything was all right. I would try to pick her up and bring her to bed, but usually she would not stay. She would go quietly back to the Bucket and not cry out again that night. But she might the next; you never knew. We joked, uncomfortably, about her "arias," because she had begun trying out different sounds, many of which we had never heard before, all of them loud. She had never been loud.

If we had known then what the veterinarian Jessica Remitz has recently written about that phenomenon, we'd have been more upset—but, as is unhappily so often the case with aging cats, we wouldn't have been able to do a thing about it: "Cats that may be experiencing cognitive issues (such as early Alzheimer's or dementia) are very vocal during the evening and will meow as if they are lost."[10]

Well, I could have said to Dr. Remitz, if Augusta was, in fact, suffering from dementia, she was doing so only in isolated episodes. Augusta and I spent a full two months of the summer of 2007 on the Sweet Grass, and for her that summer was heaven. Some days she hunted to the point of exhaustion, and would sleep in undeniable peace and comfort at the foot of the bed. Other days she was content to sit on her haunches on the porch and watch the comings and goings of our hundreds of birds. When she traveled into the cottonwoods, well out of sight of the house, she always knew the way home precisely, and could gallop there lickety-split if a deer barked at her or I called her to dinner. On the Sweet Grass she never cried out at night or showed any other sign of disorientation.

Midsummer for me was not so fine: Elizabeth and I and a group of friends went on a backcountry pack trip to celebrate my birthday, and my horse reared and then fell backwards almost on top of me. If he had landed true, I would be dead. In any case several ribs were broken and my sacroiliac wrenched, but I was alive.

True to form, as I lay in bed for the next couple of weeks, Augusta stayed with me. I believe she knew I was in pain. Did she know I knew that she was, too?

In December 2007, at what my calendar recorded as a "routine office visit" to Pets Unlimited—not with Randy Bowman this time—Augusta's weight was down to 7.4 pounds. Neither we nor the veterinarian put together the fact—hideous in retrospect—that that was a loss of 1.7 pounds since the previous December, 18 percent of her weight. Worse by far, now that I was doing the arithmetic, I realized that over the course of the last two years she had lost 2.6 pounds, *26 percent* of her customary weight. Why had none of us, why, especially, had none of the *doctors* at this top-flight hospital, taken adequate notice of such a drastic decline?

We really had all believed that we were doing the best we could for her. The best thing was that her pain had been much relieved. If she lost weight, was that so bad? It meant less strain on her joints, I tried to tell myself.

I spent June 2008 alone at the Sweet Grass place, and then Augusta and Elizabeth joined me for July. When you're with your cat every day you may not notice change as it comes slowly on. When Augusta arrived in Montana after we'd been apart for a month, her appearance shocked me. Despite a five-milligram shot of prednisolone only a few days before, she looked disheveled and confused. She seemed not to have been grooming herself well at all, and her vertebrae made a line of sharp bumps from between her shoulder blades to the tip of her tail. Her whiskers had been white for a couple of years, but I had failed to notice the faint graying of her soft little muzzle.

Restored to her natural home, though, she brightened quickly, as she always did, and soon she was picking her way through the grass, which had grown tall that summer after a stormy spring. Augusta's eyes shone happily. From time to time she would sproing into the air and describe the balletic arc of her youth, and a few of these ended in the actual capture of a mouse or shrew.

At the end of September 2008, Augusta returned to the vet for a "senior wellness profile": a urine sample, blood tests, a thorough poking and prodding, a renewal of her vaccine against feline distemper and rabies. Augusta had been vomiting, rather often, and the vet suggested we elevate her bowl a couple of inches so her food would be easier to swallow. Also we should watch to see if fish-based food caused more vomiting—it was often a culprit, he said.

The test results showed that her red and white blood cells were fine, her liver was normal, her calcium normal, her electrolytes normal, her triglycerides low. The hyperthyroid test was negative. Best of all, her kidneys were normal—urological problems are the curse of the aging cat. The doctor pronounced her condition "very nice for thirteen." But she weighed only 6.7 pounds, down 9 percent in nine months, a cumulative loss of 3.3 pounds. One-third of Augusta had wasted away.

If she continues to lose weight, said the doctor, the next thing to do is an ultrasound of her abdomen, to check for masses—tumors, abscesses, and such.

At her visit to Pets Unlimited in early April 2009, we saw Randy Bowman again. I told him that Augusta had been eating well for a while, but she had lost another half-pound—she was down to 6.2. She had soured on her usual Iams food, and I had been trying a bunch of highly touted, high-priced cat foods I had found online, without much success. I told him that recently Augusta was constantly asking me to "make her a drip"—that is, to set the kitchen sink faucet so that it barely dribbled, her favorite way to drink water. She was thirsty all the time. That could be a side effect of the prednisolone, Randy said. Palpating her pelvis and her rear legs, he told me that they had deteriorated significantly since the last time he had seen her. He noted that her coat was in poorer condition as well—oily and clumpy. He took her temperature, examined her eyes and teeth and mouth, checked her for swollen glands. He said that other than the weight loss, she looked pretty good—"really not geriatric."

As for the weight loss, hyperthyroidism was a possibility even though her last test for it had come back negative. If that's what it was, it was eminently treatable. Her blood sample was going to get a specialized thyroid test, and I could call for the results tomorrow. Inflammatory bowel disease was a possibility, also easily treatable; I was to bring in a stool sample. It could also be a lymphoma. We could treat that with chemo, he said, which is expensive, and unpleasant. In any case the disease is invariably fatal.

Her thyroid and IBD tests both came back negative. We managed, somehow, to avoid the unavoidable inference, that she had cancer. If we had recognized it, would we have asked for an ultrasound? And if a lymphoma had been diagnosed, what would we have done then? Knowing that it was terminal, I'm certain we'd have done nothing more than what we were already doing. As long as Augusta was enjoying her life—which she indubitably was—the best we could do was to keep it as enjoyable as possible.

At Randy Bowman's suggestion, we tried a number of cheap mass-market cat foods, and Augusta seemed to like them, but only for a short time. Then we discovered Fancy Feast, which we just called "white food" because every flavor came drenched in some sort of creamy-looking white goop. A better name might have been Kitty Crack, because Augusta couldn't get enough of it.

She began to gain weight, and her midnight arias waned. By May 2010 Augusta had made it back up to seven pounds. In July we took her to Pets Unlimited to ask if she was okay to fly to Montana. After an extensive exam, the vet gave Augusta a health certificate. Her weight was up again, a little, to 7.1 pounds. Elizabeth thought Augusta's weight gain might have been due to the white food. I now think it was the growth of her tumor.

For the first time ever, Augusta did not do well in Montana. Her coat did not grow sleek nor her eyes bright. She did try a little hunting, but she seemed to grow disoriented, and would turn and head home. She threw up repeatedly. She did like sitting on the porch in the sun. Elizabeth and I decided that rather than cram her into the cramped confines of the air travel carrying case, we would drive home together.

At home in San Francisco, Augusta grew weaker by the day through the month of August. She turned fifteen years old.

♕

Because veterinary medicine is so good these days, cats are living longer, and advancing age brings increased vulnerability to accidents, infection, cancer, parasites, and a host of afflictions that probably weren't troublesome when your cat was younger. Many of them are eminently treatable now, but as a cat ages, illnesses that were previously less serious may threaten to be fatal, and damage or disease that will inevitably end in death becomes more likely. Advances in veterinary research and practice, however, have now made it possible that in some of those terminal cases, the end can be postponed.

The question of how far to take medical treatment of pets is a hot one in the veterinary world. In recent years there has been a dramatic increase in specialization and in the use of technologies previously reserved to human use. As among our own physicians, specialists are highly motivated to treat their specialties, and just as we have to look to our own real interests in the medical world, you cannot necessarily accept that every word from a veterinarian's mouth is going to be appropriate to your and your cat's situation. Veterinary oncologists, for instance, make their livings treating animals with cancer, and they now have at their disposal veterinary chemotherapy, radiation, and nearly all the other weapons developed for human oncology.

Picture yourself in this situation. You're told that your cat has cancer, and that chemotherapy might extend his life by six months or a year. You're aware that chemotherapy usually exacts its own high cost in misery. But you could have him with you for another year, alive, purring, loving you, and being loved. Sometimes you must look deep in your heart to find whether you want that extra year for the cat's benefit or for your own.

Veterinary oncologists do make a strong case for their ability to ease suffering. Dr. Joanne Intile writes, "When aggressive surgery is not an option because an owner feels their pet is too old to withstand the operation, veterinary oncologists are able to offer less intensive chemotherapy therapies, most often designed to slow tumor growth and metastases while maintaining an excellent quality of life. Though we may compromise our chance for a cure, we are able to extend an animal's expected lifespan and simultaneously ensure that their remaining time is spent as happy and healthy as possible."[11] Because cats' happiness or absence of it can be so difficult to read, the value of such therapy can be difficult to judge.

There are always just enough *rare* cases of miraculous cures to keep countless flickers of hope alive. It's extremely hard to think about statistics when you're holding your beloved companion in your lap and she's very ill. When you're facing the prospect of many thousands of dollars in medical bills, there can be a curious psychological reversal. You wonder if, if you withdraw treatment now, will you be wracked with guilt for the rest of your life? You wonder, Will I be forever hearing a voice inside me whispering, *I killed her to save money?*

"Many people are enrolling their pets in insurance programs to help in the case of catastrophic illness," writes veterinarian Jessica Vogelsang. "It's the most likely way to save a life down the road." *But* "It is not too uncommon to hear of bills over forty thousand dollars."[12]

End-of-life decisions were easier not long ago. The adage was,

She'll tell you when it's time. You'll just know. When it's clear that your cat no longer enjoys living, or that more of the things she has always loved to do are now beyond her capability than the few things left she still can do, you know. She stops grooming herself. She doesn't seek out her favorite spot in the sun. She won't eat. She stumbles. Even when she lies down you can tell she can't relax—because she's in pain. Then you know.

Or you used to know. Because now perhaps steroids, or chemo, or some other advanced life-lengthening treatment may be available, and you're going to have to *guess* whether your cat's lengthened life is likely to be worth living. Is there a realistic prospect of a cure? Could six months of chemo buy you three healthy years? It won't be easy to decide how much confidence you have in a given doctor's answer. With years of experience together and gut-level trust in your longtime vet, it may not be so hard, but you're likely also to be dealing with a specialist whom you don't know and who doesn't make any money by recommending against treatment. In any situation in which the cat's life is at risk, a second opinion would seem a worthwhile next step.

Asking your vet for a referral can be difficult. Vets are human beings. You may feel you're somehow insulting them, questioning their judgment. Don't. Referrals for second opinions are part of their lives. They're trained to do it, and they can learn from it. And that second opinion could save your cat's life.

A relatively new development for pets nearing the end of their lives is the animal hospice, which is quite similar to a human hospice. It is rarely if ever a place; it is a system of care that looks to all the needs of an animal near death, to an extent that may be beyond the capability of the pet's owner—pain management, nutrition, hydration, hygiene, physical comfort. A veterinarian named Alice Villalobos has developed an evidence-based "quality of life scale" to help hospice workers in their decision making. It may be of help as well to some cat owners stymied by indecision, independent of any hospice. It's still not easy, but by breaking down the decision about ending your cat's life into meaningful components and giving them a numerical score, it may sharpen your thinking. Each category is to be scored from 0 to 10, with 10 being ideal:

_____0–10 HURT. Adequate pain control, including breathing ability, is the first and foremost consideration. Is the cat's pain successfully managed? Is oxygen necessary?

_____0–10 HUNGER. Is the cat eating enough? Does hand feeding help? Does the patient require a feeding tube?

_____0–10 HYDRATION. Is the patient dehydrated? For cats not drinking or eating foods containing enough water, use subcutaneous fluids once or twice daily to supplement fluid intake.

_____0–10 HYGIENE. The patient should be kept brushed and cleaned. This is paramount for cats with oral cancer. Check the body for soiling after elimination. Avoid pressure sores and keep all wounds clean.

_____0–10 HAPPINESS. Does the cat express joy and interest? Is the cat responsive to things around him (family, toys, etc)? Does the cat purr when scratched or petted? Is the cat depressed, lonely, anxious, bored, afraid? Can the cat's bed be near the kitchen and moved near family activities so as not to be isolated?

_____0–10 MOBILITY. Can the cat get up without help? Is the cat having seizures or stumbling? Some caregivers feel euthanasia is preferable to a definitive surgery, yet cats are resilient. Cats with limited mobility may still be alert and responsive and can have a good quality of life if the family is committed to providing quality care.

_____0–10 MORE GOOD DAYS THAN BAD. When bad days outnumber good days, quality of life for the dying cat might be too compromised. When a healthy human-animal bond is no longer possible, caregivers must be made aware that their duty is to protect their cat from pain by making the final call for euthanasia. The decision needs to be made if the cat has unresponsive suffering. If death comes peacefully and painlessly at home, that is okay.

_____TOTAL. A total score over thirty-five is acceptable quality of life for maintaining a good feline hospice.[13]

Hospice care can be a compassionate alternative for cat owners who simply aren't ready to turn out the light, but whose cats are suffering and are never going to get better. There are now a number of organizations that can help you find a good hospice, among them the Animal Hospice End-of-Life and Palliative Care Project, the International Association of Animal Hospice and Palliative Care, and the American Association of Feline Practitioners. All can easily be found on the web.

The unbearable decisions pile on without mercy. Some people elect not to make the final decision, but to let nature take its course. Sometimes, when you haven't yet decided how to decide, nature will sneak up behind you: You go to check on your cat and he's already dead.

Most cat owners eventually accept the choice of euthanasia. The American Veterinary Medical Association has gone into the philosophy and practice of euthanasia in all its dimensions. The AVMA guidelines, in 102 compassionately considered pages, range from definitions of "consciousness and unconsciousness," "pain and its perception," and "stress and distress" to "a good death as a matter of humane disposition" and "a good death as a matter of humane technique" to rules for acceptable and unacceptable methods of euthanasia and "behavior in presence of owners."[14]

A generation ago, veterinarians were much less sensitive than most are today. It cannot be easy to kill innocent, uncomprehending animals day after day and to bring compassion to their sorrow-sickened companions. To seal out their own emotions, no doubt often both strong and mixed, and to wall themselves off from their clients' grief—as so many practitioners of human medicine do—would be perhaps a less draining course, at least in the short term. The decency and the concern that we see in our veterinarians are worth remembering to be thankful for. Bringing the mind to rest on gratitude to someone else is not easy in a time of such entire focus on one's own emotional state.

That some vets now will even come to your home to end your cat's life seems to me beyond remarkable. Imagine being that veterinarian, driving or on the train, black bag in hand, on that errand of unspeakable mercy. What, if any, music can they play? What, if any, thoughts can they allow, and to which must they deny entry? Now that there are a few places where human euthanasia is permitted, the same but perhaps even more heart-choking agony must occupy those needle-bearers' rides. Yet maybe no: For the human will have chosen to die and will be facing death in some degree of courage and understanding. We call it "assisted suicide," aptly. The physician's burden is, morally and philosophically, that of an assistant, albeit a brave and admirable one. The veterinarian's burden is entire, private, and unimaginable. Until the very last, the cat in almost every instance looks into your eyes in unknowing innocence. The hand that pushes the plunger down the hypodermic shaft belongs to a person who has chosen to undergo an extraordinary suffering to render an extraordinary kindness.

You will have some moments to recognize as your last time

together. You may choose to be present at your cat's death, or not to be. Hard as it is, it seems better for most people to be there at the end. You can hold him if you wish. There will be an injection of a sedative, and he will drift softly into unconsciousness. Then, after a second injection, into a vein, he will take a deep breath, perhaps another, and fade away.

If you have a burial place, you may take the body away. If you have chosen cremation, the body will be taken now.

ℐ

It will tear a hole in your life. Her love was unconditional. When you stayed away too long, she didn't sulk when you came home, she welcomed you with gladness. She was so innocent. So naïve. No human being ever loved you with the purity of her love.

Did you tell her things you never told anyone else? Did she purr just because you were there—because you *existed*?

His stuff is going to be all over your house. What are you going to do with his bed? His toys? You're going to listen and listen for the bup-bup-bup of his paws on the floor as he comes trotting to greet you, and you won't hear it. You're not going to be able to sleep. You're going to eat too much, or not enough.

You're going to wonder if there's something wrong with you. A lot of people stay home just so they won't have to hear somebody say, "Come on, *it was only a cat*."

Somebody's going to tell you to get a kitten, and you're going to think, No! No kitten could possibly replace her. You're right— don't get a kitten until you want one. You'd only be making invidious comparisons. One study, in Scotland—with dogs, so we can't be certain whether it would apply to cat households—showed that getting a puppy when your old dog was still alive but dying could ease the pain, especially if there are kids in the house.[15]

The children are going to be hysterical with grief in any case, and then, after a couple of days, maybe one of them isn't going to care at all. Younger children seem to be more resilient, but if your cat's life ended in euthanasia, chances are that children of any age will blame you.

You may be blaming yourself as well.[16] You may be secretly comparing the intensity of the grief you're feeling now with the grief you've felt in the past for some person very close to you and realizing that the loss of your cat hurts more. You may blame yourself for that too. You may find comfort in the knowledge that serious studies have shown that grief for the death of a pet and grief for the death of a person are psychologically indistinguishable.[17] No comfort? There may be no comfort to be found now, and there may not be any comfort for a while.

Your other pets, if you have any, will grieve, often deeply. A surviving cat will probably search for his lost companion, returning and returning to the places they shared, sniffing for now-absent scent. He may produce sounds you've never heard before, yowls from deep in his throat, calling his friend to come home. He may stare blankly out the window. Like you, he may not be able to sleep or eat. He may cling to you for comfort, or try to soothe himself by overgrooming. His grief, like yours, may not diminish quickly. Some people believe that letting him see the body just after his companion's death may ease the pain; others consider that absurd—there has been no research on the question. What can you do? Attention, kindness, love may help to fill the emptiness, and comforting the grieving animal may comfort you too.

And what if another person in the household isn't grieving, doesn't care? Horrible. Divorce rates as high as 23 percent have been reported for bereaved pet owners.[18]

Take a deep breath. Not ready? Don't worry. You know you're

going to have to take yourself in hand sooner or later. You've known it all along. All these things, you can do them through tears.

First, memories. In honoring your memories of your cat, you are honoring him. Try to remember everything. The order isn't important.

You felt his warmth and his breathing, and sometimes you would slide your hand beneath his chest to feel the amazing flutter of his heart. When you touched his fur, it was alive, it rose to your touch, and then as it relaxed you could feel the silent sigh. Remember how he turned his ears toward the front door before you knew someone was coming. There will be no end to these memories, and let there be no end. Write them down, or speak them into a recorder. Keep them alive.

Dabbing in the sink with her paw for a leaf. Waiting till you had established the fine and regular drip—*bip bip bip bip bip*—then swatting at it till she was ready to turn her head sideways and lap from the thread-thin stream. The wind ruffling her fur from behind revealing patterns you had never seen—dark brown and darker brown tabby stripes. Our little black cat wasn't black! The twitching of her legs and lips and the squinching of her eyes in a dream—of what? How she would push, push with her back feet against your hand, the cool rabbit-soft fur of her heels, the lizard-skin one-way roughness of her hot toe pads. The little pocket on the side of her ear, lined with ineffably soft fur. Was that little spot perhaps a refinement of her hearing apparatus? What matters now is that it was a place which in perfect trust she loved your finger to caress from within. With the same trust, how she loved you to press your finger to the inner point of her closed eyes and run it softly outward across the lid, perhaps picking up along the way a bit of oozy gradu. Her pink tongue when she gave herself a bath.

How she always knew the sound of you coming up the front steps, and always would come trotting to greet you with tail held high.

I remembered. Augusta sniffing my extended index finger first thing in the morning with great interest, almost as if it was new. Augusta plunging into laundry warm from the dryer. Augusta jumping up on the dining table and biting the flowers, preferably tulips. She liked for us to watch her when she used her litter box. She liked when we threw out the old sand, washed the box, and filled it deep with new, unscented litter. She would wait, transfixed, excited, and before the dust had settled she would have jumped in, dug like a fiend, and christened it with a big fresh poop, eyes shut tight in pleasure.

All the times when she walked away in indifference. Times when I called, *Augusta, Augusta!* and she would not come, even when I tracked her to her hiding place and I insisted, *Augusta, come down!* but all she would do was tiptoe back and forth on the branch or in the attic corner or wherever the hell she'd gotten to, in some impenetrable trance, mewing pitifully, not even looking at me yet pleading *Please help me!* still refusing the slightest recognition of my exhausted repetitions of *Augusta, come!* till at last all I could do from the teetering ladder top was seize her by the scruff of her skinny little neck, and even as I rescued her she would bite me and scratch me till I clasped her to my chest not lovingly but like a firefighter pulling a crazed child through flames. These and a hundred other memories I wrote and wrote. I made a list of her names, Bdingle, Bdomble, Busta, Doodoo Head, Dummy, Little One, Panterina, Piece of Shit, Schbdingle, Schkblodgit, Stoopie, You Idiot, Beauty Kitty. I kept remembering more. Augusta? *Augusta!*
Grieving. Helpless.

⁂

In September of the year 2010 Augusta began to eat less and less. She acted hungry, but then would take only a few bites. Her affect was foggy, absent. She moved very slowly. She had lost another 20 percent of her weight in six weeks. The veterinarian she had always loved, Randy Bowman, found a hard mass in her lower abdomen. An X-ray confirmed that it was a large tumor, probably a lymphoma. She was going to die.

Randy gave us pain medication to spread on her gums, and an appetite stimulant, but she could not hold the medicine down. She began to hide in Elizabeth's closet. When I looked in, she looked back and me and cried in a way I had never heard before, a low moan. When she came downstairs, we tried to think she was better. We tried all her favorite foods. She sniffed almost eagerly at the cantaloupe, but she would not eat it. I put out a few of her favorite crunchy treats, but she seemed unable to find them on the plate. She went back to the back of Elizabeth's closet. Elizabeth spent that night in a sleeping bag just outside the closet, to be with Augusta.

In the morning, Augusta roused herself and went downstairs. She ate a few bites of white food, used her litter box, and lay down in the sun on the kitchen rug. She brightened, ran upstairs, not ready, after all, to die that day. We decided that if she was better again tomorrow, we would wait and see if she might start eating and feeling a little better; that if she was worse, we would go ahead and have her put to sleep; and that if she was the same, we would have to decide then.

I spent most of that night in the sleeping bag next to Augusta. She left the closet at least once, when I was asleep, to eat some of the raw hamburger that we had left out near the closet and also to use the litter box in the bathroom across the hall. In the morning she was wide awake but showed no inclination to come out.

She did not sleep at all. Her eyes were open, her face a blank, an absence. She was purring constantly. I had read somewhere that a constant purr means that the cat both knows she is dying and is comfortable.

Elizabeth and I agreed that today was the day. Augusta would face only further weight loss, and further decline. It would not be long before some vital organ would fail. In recent days she had seemed no longer to experience pleasure beyond a few seconds of petting or brushing—and even that soon became bothersome and she moved away from it.

We called Pets Unlimited and made the appointment with Randy Bowman. Augusta lay curled up well back in the closet, in the dark, purring. Occasionally she changed position very slowly and, I thought, painfully. Occasionally I stroked her head, and I thought she liked it. She began to lick a front paw, as though she was going to wash her face, but then she laid her chin on the paw. From time to time she looked at me. From time to time she closed her eyes. Mostly she just stared at nothing.

At 12:40 we were shown into a cold examining room at Pets Unlimited, and Randy came in behind us. We put Augusta on her own blankie underlain by a pad on a cold metal table. She was calm. Randy explained that he would first give her an intramuscular injection that would act over the course of five to ten minutes to sedate her into unconsciousness. He administered that at 12:45, and it hurt—Augusta squirmed and turned as though to try to bite him. She made brief eye contact with Elizabeth, and then she relaxed quickly, staring straight ahead, very still. I watch her flanks moving as she breathed, and as her breath slowed very slightly. As imperceptibly as the hour hand on a clock, it seemed, she lowered her head to the towel. We continued to stroke her gently as she relaxed, relaxed.

By 12:50 her nose was on the towel. I lifted her chin so she could breathe more easily. Randy checked her blink response, which was still there. Two minutes later it was not. He took an amazingly noisy electric shaver to the front of her right front leg, making the vein there easily visible. She did not react at all to the noise, which under ordinary circumstances would have scared her. Her eyes were open, but she was unconscious.

Elizabeth and I both continued to stroke and hold Augusta. Elizabeth asked Randy to show her where to put her hand so she could feel Augusta's heart beating. At 12:55 he injected a large syringeful of barbiturates into the vein in Augusta's shaven fore-leg. Her heart stopped instantly. Randy told us that her brain had died equally instantly.

We had read to expect several possibilities: a series of deep, searing last breaths; shuddering; urination; a release of her bowels. None of these came to pass. She was simply utterly still.

We stayed with her a few minutes. She looked exactly like herself alive. I tried unsuccessfully to close her eyes. I put my fingers between her toes, something she didn't like much when she was alive but something I always loved the feeling of. As we left, Randy was wrapping Augusta in a towel, to take her to a freezer.

⌁

She was never afraid of us. Inevitably sometimes we would step on her tail or trip over her, but that left no memory: She was never afraid we would hurt her.

Soon her places were empty, the litter boxes gone, her food bowls, the Bucket. The pantry shelves where we kept her food were bare. The library steps we had gotten so she could climb up onto our bed were in the basement, awaiting the Salvation Army.

Up until her last morning, she would come when I called, even up or down the stairs, "Augusta, come!" In my mind I couldn't stop calling her name.

I collected her toys, her blankie, her ribbons. Beside the back stoop I dug and dug until the hole was three feet deep. We took a last look at the little pine box that her ashes had come in from the crematorium. I reached down and placed it softly on the bottom of the hole. We covered that with her stuff, the Anchovy Mouse, the Spider Ball, the strip of cloth with AUGUSTA cut from the Bucket, the lists and the memories, then dirt. We planted a tall, thin Coprosma that we would see every time we went in or out of the door that Augusta went in and out of so many times.

Chapter Eight

Love

This book does not entertain the notion that cats are incapable of loving people. You can find that argument made elsewhere.

Still, think about this. What *are* we doing, keeping this animal in our house?—loving this creature, whom, no matter how hard we try, no matter how many studies we study, we ultimately cannot understand? Sometimes, when I looked at Augusta drowsing in a chair, it seemed downright *weird* that she was there at all. In our life, I mean.

If weird connotes unusual, there is nothing weird about it. Somewhere between thirty-six million and forty-three million American households have at least one cat.[1] How many American pet cats in all, then? Seventy-four to ninety-six million.[2] The United Kingdom has eight million pet cats, France ten million. Italy, with a human population of sixty million, has nine million cats (that's just the pets, excluding their millions of ferals).[3]

So it's not weird to have a cat. One still sort of wonders why we do.

The psychiatrist Aaron Katcher maintains that pets "help preserve our mental and physical equilibrium," and he has identified four kinds of "transaction" between us and our animal companions that bind us together deep in our beings: safety, intimacy, kinship, and constancy.[4] Katcher's signifiers are worth looking at one by one.

You can tell that Katcher is more of a dog guy when one of his first examples of the safety principle is that you can talk more easily to a stranger out walking a dog than to one who is out for a walk without a dog. The same, he observes, goes for someone out walking with a baby. Not many people walk their cats. What's more, "Considering the almost universal proscriptions against touching strangers, it is important to note that one can touch the child or the dog as a means of greeting." Even if somebody had succeeded in walking a cat, a head-on pet-the-cat approach would be a dubious proposition, though it might be a good conversation starter (unless the walker was batty, not entirely improbable). Nevertheless, when Katcher writes that people in "dyadic relationships"—he means couples—sometimes find that they can discuss difficult topics in a kind of "triangulation" by talking first *to* their pet and eventually *through* it, a cat comes readily to mind. "Animals," he says, "make people safe for people."

Victoria Voith, whose work we met earlier, writes insightfully about this sense of safety: "Human beings evolved in social groups that lived in close proximity with other animals.... The animals serve as mutually reciprocal sentinels for danger.... As long as the animals around us appear serene, 'all is well.' We can decrease our vigilance. Perhaps the presence of nonfrightened animals sends a message, critical in our evolutionary past, that there are no predators in the vicinity and it is safe."[5]

Intimacy, by definition, says Katcher, combines talk and soft touch. Many people not only talk to their cats but confide in them. Just how many of us think the cats understand what we're saying doesn't seem to have been determined, but 92 percent of Voith's study subjects believed their cats were at least attuned to their moods. "Pets," she writes, "can generate a feeling of well-being, a feeling of being loved. Pets appear to miss us and to be happy to see us. They seek us out to be touched and, perhaps more importantly, to touch us and provide soft, warm, tactile contact with a living being...."

Again focusing on dogs, Katcher made observations of intimacy most of which also apply accurately to cats: "The person directs his or her gaze at the dog.... The dog is stroked and talked to.... The person's voice becomes softer.... The cadence of the owner's speech changes, and fewer words are uttered per minute. A pseudo-dialogue is established by alternating questions and silences.... Blood pressure is lower.... There is a change in facial expression... a smoothing of the facial features, with the loss of signs of tension and the presence of a fixed smiling...."

Quoting Katcher is a joy, for the sheer bloopy formality of his language—"dyadic relationships," "a fixed smiling." And now the *ne plus ultra*: "The presence of a dog in an experimental chamber, a veterinary clinic, or a home permits the study of intimacy without sex in the same way that the procedures of Masters and Johnson permitted the study of sex without intimacy." Zing!

Aaron Katcher's third "outcome of transactions between man and animal"—kinship—at first struck me as another eccentric choice of words. Then I remembered that 99 percent of Victoria Voith's subjects considered their cats to be members of the family. That pretty much sums it up. Katcher presents an amusing sequence of examples, all dog, though you can substitute "cat" and they still work: "Family portraits when taken in a setting about the home tend to include the [cat]. The [cat] shares the parental bed and bedroom. The [cat] is fed and does not feed himself. The [cat] is permitted to display his anus and genitalia. His excretory functions are a subject of public concern, effort, and direct comment without embarrassment. He is talked to but is not expected to use words.... The manner of speech used toward the [cat] frequently resembles that used with small children...[the] lilting, simplified language of 'motherese.'"

For Katcher's fourth essential quality of animal companionship—constancy—he draws distinctions between children and pets. Again I substitute cat for dog: "Since he is not expected to become adult, there is no demand that the [cat] progress along an axis of intellectual, moral, or social achievement. The [cat] stays the same. He never grows up, never learns to talk, or how to care for himself, or how to wear clothes that hide his genitalia. He never learns shame."

Katcher also makes a couple of assertions about what he considers the *proper* nature of having a pet. "The tendency of children to make partial objects out of animals, by attributing to them extreme qualities of virtue or vice," seems to Katcher to infect some grown-up pet owners. "This way of animating and personifying animals can fail to recognize the existence of the real animal. Moreover, it is frequently an expression of a highly narcissistic attachment, in which the affection and love attributed to the animal is, in reality, affection directed at the self....Like the self, they are highly individualized, as stuffed animals or security blankets are individualized....People who love animals in this fashion are more vulnerable to the loss of the animal and can experience intense depression following such loss."

This stung. Was the depth of my grief over Augusta "an expression of a highly narcissistic attachment"? Was she no more than a projection of what I wished her to be? Then I read Katcher's rosy alternative:

"There is, however, another way of being with animals....It can recognize the reality of animals without split perception or narcissistic love. It recognizes the identity of the animal with his species....This is the kind of relationship with animals that is shared by the farmer, and even by the hunter....We have lost the notion that it is possible to love animals and see them die or sacrifice them in their time....With a change in vision, and a change in position, man can stand outside of the individual cycle of any one dog and see the repetition of that cycle many times in his own existence."

Whoa now, Aaron! I'm supposed to think of Augusta as *livestock*? Or as the hunter sees the deer—there'll always be another one?

After a few deep breaths, I began to see some truth in both of Katcher's extreme ideas. Of *course* we project our own needs onto the pets we love—we project our own needs onto the people we love, too. We project our own needs onto the movie stars and the works of art and the landscapes we love. And yes, yes, yes, we must remember—I guess I've been trying to make the point throughout this book—that these are animals with selves of their own, selves we can see into only dimly but must respect, and must try, rationally, to understand. In turn they will try to teach us the truth of themselves.

Not being much of a cat video fan, I tend to forget how many people find so many ways to turn their poor cats into something else. Are they making fools of the cats? No, because by the grace of God cats don't seem to be equipped with the faculty of embarrassment. Are those people making fools of themselves? Possibly not that either, because to make a fool of yourself you've got to realize you've done it. I don't get why somebody needs to dress a cat up as Santa Claus or a clown for the cat to be entertaining. Cats just *are* entertaining. And this is where I add a fifth essential element of the human-and-animal "transaction" to those proposed by Aaron Katcher: entertainment. They entertain us. And if we're living up to our end of the bargain, we entertain them.

It's a mutual obligation. Elizabeth was astonished when as soon as Augusta arrived in our life I said, "You have to play with her." Elizabeth didn't cotton to being told by anybody that she *had* to do anything, but it didn't take long for her to realize that she would be richly recompensed. Is there anything on this earth more entertaining than a kitten? I suppose I'm easily amused, but I believe that for every minute I spent amusing that cat she amused me back fiftyfold. When she tootled along the upstairs banister with a twenty-foot free fall below her, then drew all four little black feet tight together and flew six horizontal feet to the bookcase, that was just…Supercat-cool. When she tangled herself up in a ribbon in the legs and stretchers of a chair and exploded out of it like Houdini, I thought it was a riot. When she put her front paws up on the rim of the toilet to watch the stream of my pee hit the water, we both found that most entertaining. When she sneaked onto the dining table and started trying to chew up the tulip leaves just to bother us, that was obnoxious, exactly as she intended, but also, as she probably didn't intend, funny. I have a little rubber monkey that hangs by an elastic string from a file rack on my work table; when Augusta was hungry or bored, she would thrrrum the string and make the rubber monkey bounce, shattering my concentration. I know she thought that was funny. We used to play one of the dumbest games ever invented: She would sit on the kitchen counter and I would put a ballpoint pen beside her and she would knock it onto the floor; I would put it back; she would knock it down again; put it back, knock it down, put it back, knock it down, again and again—I don't know how to tell you how much pleasure this gave both of us. I liked to watch her sleep, and when she cracked an eye to check that I was still there, I'm certain she liked that I was watching her sleep.

From consideration of the empathic pleasure that mutual entertainment provides, a sixth essential element of the human–cat relation begins to arise: love. We love our cats with a purity and grace not possible in our love of our spouses, our parents, or even our children. People are too complicated for love as simple as what we bear to our cats. It is not *agápē, philēo, caritas, amor,* or *eros* (roughly: selfless love, brotherly love, love of humankind, romantic love, and erotic love, respectively). I believe it has never been named. The kinship between our cats and ourselves reaches deep beneath consciousness, to a place before history, perhaps even before the development of the self-expressible human intellect. In its purity our love of our cat resembles an infant's adoration of his or her mother, but it lacks that utter dependency. We do not want to relinquish it, God forbid! but it is nonetheless voluntary. It is devotional, like prayer, and like prayer it is met with silence. Our devotion is what gives cats their power.

There is a famous story in Islam that one day the prophet Muhammad's favorite cat, Muezza, was asleep on the sleeve of his robe when the call to prayers came, and rather than disturb Muezza, Muhammad cut off the sleeve. When the Prophet returned, Muezza awoke and bowed down to him. (Cats do that.) Muhammad stroked Muezza's head three times, granting to all cats seven lives and the ability always to land on their feet.[6]

If we need proof more tangible than what we know in our souls—perhaps more useful if you want to undertake arguments with doubters—there has been a good deal of research into the neurochemistry of attachment and love, showing, among other things, that one of the consistent markers of emotional attachment is an increase in the release of the hormone oxytocin by the hypothalamus.[7] This has been familiar in the study of the human brain for years, and more recent research has shown that the same thing happens in the brains of cats in affectionate interaction with their human companions.[8]

The great psychologist Boris M. Levinson tried in the 1960s to establish that the depth of awareness between animals and children could be the basis for effective psychotherapy. At a meeting of the American Psychological Association, he was laughed at. The National Institutes of Mental Health turned down his application for funding. Twenty years later, however, he was able to write that "The field of animal–human relationships is now respected as a legitimate area of scientific investigation, [although] it is not yet a full-fledged discipline." It was well on its way, in any case, and Levinson felt confident enough to say, "We may now be ready to translate into reality the myth of the golden age when animals and humans lived at peace with each other."[9] (He did not live to see it: In 1984, at the age of seventy-six, while at work in Brooklyn at the Blueberry Treatment Center for Seriously Disturbed Children, Levinson was felled by a heart attack.[10])

Levinson's signal contribution was his insistence that scientific methodology alone could never succeed in comprehending the complexity of the relationships between people and their animal companions. Research, he believed, must always allow for the "intuitive (the folk way) of studying the animal, the way used by the artist, poet, writer, plain people for generations," because

> our early ancestor regarded animals as rational beings and as partners in life....There was an understanding of how an animal felt and a corresponding respect for the animal's feelings and drives. Our early ancestors thought of animals as friends, knowing that, like themselves, animals loved, hated, grieved, and were supportive of each other. Early man thought of animals as having intimate thoughts and aspirations and also unseen powers and connections with nature that he did not possess.

"We were programmed to cooperate and live with animals," writes the psychiatrist James A. Knight. "Today, we do not always see nature and animals as friends and partners, and this perceptual failure may be contributing to the widespread sense of alienation in our society.....The first gods of humans were animals, and these animal gods symbolized the elemental forces of nature....

In many ways, humans have not changed over the last million years, and at some level of their awareness they still have a need to associate with animals....This is one of the universals of the human psyche, and we see this universality in our dreams, folk tales, drawings, and relationships with animals. The vestiges that humans carry of their physical and psychological past break through at unguarded or strategic moments and are expressed in their symbols and myths.

To understand ourselves, we must understand the meaning of the animal in our conscious and unconscious minds and what influence this meaning had on the development of humankind....The acceptance of the "animal soul" is the condition for wholeness and a fully lived life.[11]

In a world, and a life, characterized by constant change, by the craving for novelty, by anxious anticipation, the cat never changes. She is nature incarnate, in our house. In our heart.

Then our cats leave us and we are alone, in black emptiness. Sometimes you are divorced and the cat goes with the other. Sometimes they die in an accident. Rarely but not never, someone kills them on purpose. Sometimes they get sick and die young. Sometimes they just disappear. Most often they grow old, grow weak, perhaps contract an illness or a disability and are in pain, and you have them put to sleep. Sometimes you die first, and someone else takes them in, or doesn't.

For those who can't bear their cat's death, there is faith in an afterlife. Someone, in the last twenty years or so, whose identity is still unknown, invented a transitional paradise for dead pets called the Rainbow Bridge, where they live again, happily, while waiting for their human companions to follow and pick them up so that they can move on then, together, to heaven. The Rainbow Bridge is most often encountered in the form of a poem, which, lacking an author and therefore a copyright, has spawned quite a few books. It can be seen in a number of forms as well on the internet. Here is an example: www.indigo.org/rainbowbridge_ver2.html.

For those of us less faithful, what survives our loss is only emptiness, not a void but a wound, a piece of you ripped out and not healing. The poet Christian Wiman writes—in this case of the loss of a human loved one, but in our case, of the death of a beloved cat, equally apt—

> Somehow, even deep within extreme grief, the worst pain is knowing that your pain will pass....Consequently, many people fight hard to keep their wound fresh, for in the wound, at least, is the loss, and in the loss the life you shared. Or so it seems. In truth the life you shared, because it *was* shared, was marked by joy, by light....Excessive grief, the kind that paralyzes a person...does not honor the love that is its origin....[The] dead inhabit us—for good, if we will let them do that, which means, paradoxically, letting them go.[12]

All very well to say, and surely true in time, but you cannot force it. Augusta's death paralyzed me for two months at least, and I'm not ashamed of that. Excessiveness is in the eye of the judge, and in this matter only I can judge. I did find myself defending myself sometimes, sometimes against myself, more often against someone mystified, whose mystification as time passed shaded into annoyance, or just distance. You are as alone as ever you have been. You hoard your grief.

You stay home. Friends ask you out to dinner, you find a lie for declining. Work? You can't focus on anything, except this one thing. In the loss is the life you shared, Wiman says. Now you're supposed to see the joy and the light in it, somehow to be *in* that life. How is that supposed to be possible? The fucking cat is dead.

Support groups—new friends. Old friends. Counselors. Books—Amazon shows eight thousand books on grief and bereavement, 13,612 on death and grief. There's even a coloring book that claims solid scientific proof of efficacy.[13] The problem is, you have to summon up the energy to buy it, and the colored pencils, and the sharpener. Then you have to find the further energy to face the thing. The problem is there's a hole in your heart.

Then slowly, as slow as an arctic aurora dimmed by ice-fog—and by disbelief—a thought takes shape, renascent from Wiman: *In the loss is the life you shared.* Slowly, the absence becomes a presence. This pain is the presence of love deprived of its object. Love with empty heart, empty hands. Love can't exist alone—of course it's crying out. Love is not an intransitive verb.

You can't have Augusta back, but you could have:

A kitten.

Could you not love a kitten? Would that love not at least resemble your love of Augusta? Would it not be better than this emptiness, this voiceless yelling from the bottom of a pit? This waiting, tired, hurting, voided love is a particular love, the love of a cat, a cat gone forever but nonetheless cat-love, specific—yes, Dr. Katcher, all right—specific to the species. You can't satisfy it with a new car or a drug or a lover or even a baby. You have to have a cat.

Is this a resolution, then? No, it's only an impulse. But add this thought: There are too many kittens in shelters, there are thousands of cats killed every day because nobody has adopted them. Some shelters turn away kittens when they're over capacity. Maybe that's what happened to Augusta! That might have been why her stupid, hopeless human ex-protector dumped her in the snow that night. I have always believed that Augusta had littermates who died that night in the snow.

Elizabeth and I began to talk about it. Our new cat (what an idea, *our new cat*, O Augusta, forgive me) had to be a girl—a gentle soul, a quality rare in tomcats, I believed—and she couldn't look anything like Augusta. She had to be a kitten, so that as she grew up there would be continuous exchange of learning and investment between us. (Or was that a way of saying my inner Pygmalion wanted to be able to shape her character?) We walked back through the things we had done wrong with Augusta, and we swore we would not do them with the new cat. Most importantly, we would not leave her alone for long times.

—Are we ready?

No.

—What about now?

Not yet.

Kitten season was at its prime. (Births begin in early spring and stretch into fall, but midsummer is the peak, so twelve to fourteen weeks later will offer the widest choice of kittens just the right age.) Elizabeth sent me a link to the Pets Unlimited website. Unlimited is right. Kitten after kitten. Filters and filters: age (kitten, youngster, prime of life, oldster), gender, color, hair length, pattern. The personality types—Poet, Cat Next Door, Lionhearted—are each described in corny baby-kitty-speak, but they're probably the most important consideration.

You could shop online all you wanted, and we did, lollygagging blithely till we kept coming back to *this one*. We couldn't stop looking at her. We were doing exactly what Eileen Karsh warns everybody against—we had fallen in love with a photograph. It was time to go and meet the actual kitten.

This is where everybody needs to take a few deep breaths, or a dozen, and a day and a night—before you set foot in the place— to talk and to try to think it through. The responsibilities. The expenses. Who's going to do what. Do you really agree—has the whole household bought in. Because once you get there and something happens between you and some kitten, you can go from 90 percent certain you don't want one—you're still grieving, you were just looking—to I'm in love, this is the one, can I take her home today? in about a quarter of a second.

Pets Unlimited, fortunately, has already thought *you* through. No matter who you are, no matter what stage you're in, they've seen you before. But of course not everybody has a Pets Unlimited, and if you don't, please pay close attention to what they do, and consider putting yourself through something like the same rigors. If you do, you and your cat will be happier for the rest of your lives together.

One of the best things about dealing with an outfit like Pets Unlimited is knowing with what tenderness and expertise the cats have been made ready. Shelters like this have superb volunteers who foster kittens, sometimes motherless and very young ones. These lovely people know all the right things about handling baby kittens and exposing them gradually to a variety of people and helping them with shyness or other possible social problems. Pets Unlimited's volunteers must report precisely what they've done to build the kittens' confidence, and to minister to their medical needs, and to socialize them. They're even asked, and required to put it in writing, "What makes this kitten one of a kind?"

People considering adoption may not know how well prepared their prospective feline companions are. If you're adopting from a shelter, a good conversation with staff members who have gotten to know the kitten or cat you have in mind can be invaluable. Eileen Karsh wisely points out that all kittens are adorable, and you need to pay close attention to figure out a kitten's underlying personality type. So if you're uncertain about the temperament of the kitten you're thinking of—and you can't get good information about it—then consider adopting an adult. Grown-up cats are a little easier to figure out if you just spend some time with them. If you can find out what sort of life the cat has been living, that can help, too.

When you visit the adoption center at Pets Unlimited, a distinctive sense of well-being permeates the place, and there is a foreordained choreography to your visit. What everybody does first is take a little walk around the...I guess you have to call them apartments; they're certainly not cages. You notice that the place doesn't stink. Once in a while the dogs get to barking, but mostly they stay as quiet as the cats. There are litters of little-bitty ones with moms, still nursing. There are old ones with longing eyes. There are shy ones half-withdrawn in their little carpet-covered houses. There are pairs and trios, sometimes siblings, sometimes not, rollicking, rolling, wrestling. There are mature bonded pairs sweetly grooming each other.

Couples find each other's hands. Some people furtively wipe the corners of their eyes. Some are overwhelmed and hurry away— it's too much too soon. Few find the experience less than powerful.

Somebody opens the door. The kittens are there, and some toys. They look at you. You look at them. Out of the corner of your eye you notice that you are being watched. (You're assessing the prospective adoptees. The staff assesses the prospective adopters.)

There were three: Murray, a striking aluminum-silver tabby, named for the nerdy writer on the Mary Tyler Moore Show; Ted, the goofball of an anchorman, a darkish brown tabby; and Mary Tyler, the one we had already fallen in love with by picture alone. They were littermates. They and their mother had been found next to an irrigation ditch in the Central Valley when the kittens were tiny, and the family had been fostered by one of Pets Unlimited's saintly volunteers. All three kittens were full of beans, fearless, friendly, glad to be held, athletic, but Mary Tyler had glamour, pizzazz, star quality.

She also looked strikingly different from any other cat I'd ever seen—a riot of features that achieved nonetheless an extravagant harmony: a sleek, shiny pelt that lay flat as a leopard's, with black spots on a background of steel-silver underlain with soft tan fur; a golden sparkle at the tips of her guard hairs; a black longitudinal stripe down her back which turned into raccoon stripes up an oddly flattened, long tail; hind legs longer than her front ones; bold black Cleopatra eye liner; a broad, flat leathery nose like a lion's. All these, we learned, were characteristic of Bengal ancestry. But she was no true Bengal: Her breast and belly were fuzzy soft snow white, shading to golden fawn at the edges, and all four paws were big round snowballs with big black toepads, most unlike the delicate mince-through-the-jungle feet of the bred-to-be Bengal. Her otherwise elegantly leonine nose bore a comical splash of white on one side.

And the eyes. She looked at you, *into* you, inquiring. Could there even have been a flicker of humor? No, that's too much to assume of a cat—but play, yes, her eyes showed an eagerness for play. We picked up a little chase toy and those jackrabbit hind legs sprang her straight up, she grabbed it in both paws and with teeth too, and tumbled to earth heedless as a—kitten. Which after all is what she was. Twelve weeks.

Yes. This was the one.

You emerge into a room clean and medical, echoic, official. You sit across a cold desk from a professionally polite interviewer with a checklist of questions, the Cat Adoption Profile. This two-sided single sheet of paper starts innocently enough—name, address, etc.—but pretty soon it gets personal. *For whom are you adopting this pet?* That's an intentionally loaded question. Two of your choices are "gift" and "other." If you check one of those, you're not likely to get much further. Another choice is "family." Check that one and they're going to want to meet the family, all of them, to be sure that everybody's on the same page about wanting to adopt (often they're not). *Have you had pets in the last ten years? Dog, cat, or other? Age? Spayed/neutered?* And another loaded question: *Where are they currently?* If you left them behind when you moved, or you gave them to Granny because, oh, Granny seemed lonely, or any other answer suggesting that you might not take personal responsibility quite seriously, that's a red flag. The next question is clearly diagnostic: *Under what circumstances would you return a pet to the shelter?* The interviewer has been trained in matchmaking, and has been alert to every nuance of your answers. They can turn you down at any point for any reason.

The questions go on, and on. *How many hours a day will the household be without people? How much do you think it will cost you each month to provide the necessary medical care and to cover the costs of feeding and caring for this cat?* Many people are stunned to learn the answer: seventy-five dollars or more.

It is most important to me that my cat— (She stops abruptly in midsentence.) Go ahead, answer.

The note of forced sincerity in our voices was painful. Every answer was accurate, our desire for this kitten pure as Montana snow. So, um, could we have her?

Yes.

You sign a legal document about as charming as an apartment lease, put the $125 fee on your credit card, and—wait. She still had to be spayed, and she needed another two weeks with her mom and her sibs.

We asked if we could see her again, and they brought us and the family back to the greeting room. As soon as I sat down cross-legged, our kitten—our kitten!—padded merrily up onto my knee and looked into my face. My answer to her question was: Yes.

The other kittens held shyly back, as though they knew. Elizabeth picked up our kitten, and she curled into the curl of Elizabeth's arm and buried her face in the crook of her elbow. This was good, a good decision. We went home certain of that.

We chose to name her Isabel, not for any meaning but for its sound, its simplicity, its matching threes of vowels and consonants, its graceful dactylic meter, for how well we knew she would recognize it when we called her, how she would like it when we talked to her, how she would be comforted when we whispered it, and for the beauty of the many other ways it is spoken around the world, among them

Ysabel, Isabella, Isabelle, Isbel, Belinha, Bella, Belle, Ibbie, Issy, Izzy, Libby, Sabella, Zabel, Elisheba, Alžběta, Ella, Else, Eliška, Eli, Elise, Eliso, Lilly, Lis, Lisa, Lise, Liisu, Liisa, Liisi, Liliána, Sibéal, Ealisaid, Ishbel, Betje, Jelica, Yelyzaveta...[14]

These incantations were somehow hopeful mnemonics—we were readying ourselves for the coming days of which we knew we would want not to forget a moment. They were also hopelessly ineffective erasures of the name Augusta.

In the meantime we had homework—a corner-stapled sheaf of forty pages, printed front and back. The cover page blared
CONGRATULATIONS!
PETS UNLIMITED ADOPTION CENTER RESOURCE GUIDE FOR
NEW PARENTS
Getting Ready for Your New Cat
Bringing Your New Cat Home

Protect Your Pet from Common Household Dangers
Introducing Pets to a New Cat
Starting Out Right with Your New Cat and the Litter Box
Positive Reinforcement: Training Your Dog or Cat with Treats
    and Praise
Be Prepared: Natural Disasters and First Aid
Cat Toys and How to Use Them
Removing Pet Stains and Odors
Caring for Pets When You're Ill
Coping with Allergies
Understanding Your Kitten's Behavior and Development
Reducing Your Cat's Fearful Behavior
Solving Aggression between Family Cats
Cat Aggression toward People
Your Pregnancy and Your Cat
Preparing Pets for a New Baby
Are We All Moving?
Of Course You Can Cry

—*Of course you can cry?* Jesus, we haven't even got her yet and already they're preparing us for her death? But truly, there was stuff in here that in fifteen years with Augusta we still didn't know. Chocolate is poisonous to cats? Rubber bands can cause intestinal blockages? (Augusta played with them for years, and apparently just lucked out.) An earthquake survival kit—great idea for San Francisco! We didn't even have one for ourselves.

Before Isabel could be released, they gave us yet another stack of paper: vaccination records, need for dental care, pet insurance providers, post-adoption info sheet, thirty days of free medical care, we could "surrender" Isabel within thirty days but wouldn't get our 125 bucks back. Surrender her, my hind foot.

We went to take a look at her. She was all alone now. Our big-eyed, big-eared kitten was sporting a plastic Elizabethan collar, to keep her from gnawing at the freshly stitched incision of her spaying surgery. Could we take her home now? Could you please take off the collar?

Yes, but first, back to the big metal desk and more documents, accepting Isabel "as household pet and companion," promising to provide exercise, play, food, water, love, kind treatment, and medical aid *at once, at our own expense*, not to have her cosmetically altered or used for experimentation, never sell or abandon her, indemnify and hold harmless…any and all liability…settlements…enforceable through judicial proceedings…costs and reasonable attorneys' fees…scribble scribble. Okay, okay. *Now*?

Well, let's just run through these Take Home Instructions for Mary Tyler (their software wouldn't allow for her new, true name until five years later): heartworm—fleas—high protein low carb, canned better than dry—cats *not* respond to force—places to climb and look out of windows—room or other space her own—prevent litter box problems, get kitten off to good start—biting—

All right, *all right already!* Do they give this much paperwork to incoming freshmen at Harvard? Can we please have our kitten now?

A volunteer came out, smiling, carrying a cardboard carrying case with "Mary Tyler" written on the side.

It is a drive of seven blocks, three minutes. We have already set up the bedroom with food, water, scratching post, litter box, bed with sheepskin blankie. We carry the case upstairs, open it, and out walks Isabel. Isabel! Isabel, welcome home! She looks around. Not a trace of fear on her face, in her body. She walks forward, tail high—*I think I like it here.* She goes straight to the scratching post and plucks merrily away, proof of her good education. She looks at me, sits down, sits up, lifts her front paws high: *Pick me up!* She rests against my chest like that, white back feet perched on my arm, front feet flat on my chest, cheek to my heart—a gesture and posture that will last forever. O Love!

All the details that Pets Unlimited goes into, all their advice and demands, their lists and forms and sign-heres, may seem persnickety. I suppose they are persnickety. I wonder how many other shelters and adopters go through so much rigamarole. Is it necessary? Is it excessive? All I can say from personal experience is that they gave us a bold, calm, intelligent, friendly, non-neurotic, healthy, loving cat. For almost no money. And who but for their having taken her in would probably have grown up feral, or, equally likely, died a bad death before her first birthday.

And think. Think of the depth of responsibility we undertake when we assume one of these fragile little lives. Legally, the cat becomes our property. And morally, what? We may have to supply our own words for the obligations we incur—there may, really, be no adequate words. How many of us rush into this half-blind, and one morning, perhaps when the cat has run away, or is sick, or is acting intolerably badly and won't stop, all of a sudden we go, Oh, shit, what have I done? and we find out either that we love the poor creature more than we ever knew and we wish we had paid more attention to the details back when it would have mattered, or that we never should have adopted her in the first place and now we face the guilt and the bafflement and the brick wall of now what?

This is going to go on for years. She is going to be utterly dependent on us. Here and now: What we do in these first weeks and months will shape her behavior and our relationship all down those years. We've seen it, down through these chapters. We have to sweat the details. Isabel? Here's your litter box, here's your food, here's your blankie. Good kitty!

The light returns that we thought was gone.

❦

All over the world, cats are coming into people's lives all day, kittens, middle-aged cats, old cats, damaged cats, grouchy cats, placid cats, cats sleek as otters, bony cats, fat cats, suspicious, trusting, restless, scaredy, fuzzy, fizzy, timid, foolish, crazy, cool. Wanted, unwanted, somewhere in between. Some will be abused, some ignored, many misunderstood. Some will be lonely, some will be sad, some will run away and be run over, some will sicken and die. Some will never know a day that isn't happy, full of birds and children's chatter or timeless and serene. Nearly all will love someone.

❦

"I could have loved you better," I sang the Tom Paxton song back through time to Augusta, "Didn't mean to be unkind / You know it was the last thing on my mind." Maybe not the last thing, but bad enough. I wasn't paying attention, Augusta. How did you feel when we went away? I didn't even think. You were glad to see us when we returned, which was enough for us to fool ourselves into believing it must have been all right. *You were only a cat.*

We didn't mean to be unkind. She loved us anyway. What choice did she have? Who else was she going to love? Augusta had love inborn. She had to do something with it.

Now, bringing Isabel home, we did know better, as now you, having read this book, also know, and so Isabel and your kitten have the blessing of at least some understanding, and we have the blessing of the burden of responsibility.

Their lives are shorter than ours. We can witness their lives from beginning to end, not just witness but *be in* them, from naming to knowing, from wonder to love, an arc, and then—now—another. We can, we must mean to be kind. "Let us not be weary," wrote Paul to the Galatians, "in well doing." Every day of her life, the kitten and the cat she becomes will make the effort worth our while.

A cat shivers, wet, under a homeless addict's blanket but also is purring. He feeds her better, takes better care of her than himself. After months on the concrete they get a room—quiet, warm, secure—and the man thinks, This is the only living creature on earth who loves me.

A tribe of calicoes in a barn. A shaggy professional comfort agent in a kids' cancer ward. A skeletal ancient under the hand of a lady so old she has forgotten his name—it's a fifty-fifty bet in hospice which one will go first. A fluffy mini-Persian groomed every morning to accompany Madame in the Maybach and peek from her Birkin while she shops, has her nails and waxing done, and texts her assistant about nothing. A big-balled neighborhood bully, property of a friendless and bad-tempered exterminator who cuddles the big tom at night and murmurs baby talk to him as they fall asleep together. A mild-mannered, peace-loving house cat caught in post-divorce joint custody, one week in a gloomy man-cave with loud sports always on, then five days of children scrapping, capped by a weekend at the beach or in the snow, neither of which she has ever known or understands. A Roman ruin, cats well-fed, their every need attended to by devoted *gattare*, their bloodlines reaching back to the Empire, their late nights red in tooth and claw. A one-eyed lady cat with a permanent limp rescued from a car crash, languishing in a shelter till just short of the euthanasia deadline adopted by a timid young lady with a limp of her own. Kittens abandoned in the snow, kittens fostered in the warmest of families, kittens in thousands, in millions, learning who they are, in concert with human

companions discerning their inner lives, each teaching gentleness
to the other, all finding a new order of love.

⌀

Isabel has been trusting from the start. When she is sprawled in the
middle of the floor it never occurs to her that you might trip over
her. Still, it has taken more than a couple of years for us to find each
other's inner quiet. The other night as I sat on the sofa watching a
movie and Isabel was curled up next to me, my left hand was lying
flat when I barely felt something soft as the movement of air and
I looked down to see that Isabel had extended one paw partway
across the back of my hand. I felt like Muhammed with Muezza. I
was absolutely not going to move that hand.

Soon she was asleep, the paw withdrawn into the circle that a
sleeping cat makes. Isabel sometimes wraps her tail into the circle,
but this night she left it loose. I have discovered that I can talk to
Isabel when she is sound asleep and her tail will answer. "*Is*abel,"
I say, accenting the first syllable, and the tip of her tail flips up,
and then with two weak twitches relaxes, marking the meter of her
name.

I change the rhythm: "Isa*bel*," and sure enough, her tail replies
da-da-<u>dit</u>.

—I *love* you.

—Weak twitch, *strong* twitch, fading twitch.

—Are you *dream*ing?

—Da da *dot* da.

It's uncanny. I lay my hand on her side. She doesn't respond,
she is breathing deeply, she is deeply asleep. I can go on talking, and she will go on semaphoring back. All this is likely the
action of so-called mirror neurons, to which a number of higher
functions have been attributed and which, in that realm, have been
much argued over. But at this primal level, probably by the same

mechanism that causes people to cross their arms in sync or yawn together, mirror neurons are at work. Shall that technical knowledge lessen the mysterious bond which the phenomenon embodies, or shall the science deepen the mystery? I say the latter.

Through my hand, as she breathes, her warm flank slowly rising and subsiding, I feel the confident and tranquil love which Isabel has taught me and an awareness of a kind I can never learn.

↬

There is a cat in San Francisco who keeps a schizophrenic woman company on crowded sidewalks downtown. The woman has a box for people to put money in, and some of them, seeing how well-groomed and well-dressed she is, try to engage her in conversation. But she speaks only gibberish, volubly. The cat remains perfectly calm at all times.

Augusta, dying and in grievous pain, showed only the briefest flickers of discomfort; for years she simply lived with her illness with no diminution of her sunny contentment, and then she purred and quietly faded. Little kids pick house cats up and haul them around like floppy old Raggedy Ann dolls. While the dog whimpers under the bed, the cat sleeps through the thunderstorm.

There used to be a guy in Greenwich Village who also sat in the street, not asking for money but also with a collection box. He had a white rat that walked around the brim of his top hat and never jumped down, and a cat who was perfectly calm.

Also in the Village, at Balducci's produce market, Mr. Balducci was a surly tyrant, but his cat, though sometimes inclined not to move from his resting place on, say, the grapefruit, was as tranquil as a Tibetan monk.

When they feel safe, cats sleep, and dream. Fully awake, fully aware, they consider doing something, and sometimes they decide, on consideration, not to do it. Then they may turn to their human companions for comfort and love, sometimes for resolution of confusion, and the people feel comforted, loved, and calm. Sometimes they fall asleep together in a singularly peaceful way, in a subliminal attunement.

Is the peace that suffuses the being of a deeply sleeping cat a secret while also a constant of her inner life?

As anxiety leads to trouble, calm leads to peace. Could it be that the calm of a cat can lead us to peace?

Some cats are trained to jump through hoops of fire, and offstage are preternaturally calm. Bookstore cats are known to be especially peaceful.

Isabel asleep, the street cats, the Balducci's cat, the bookstore cats, the performers, Augusta dying, all these cats traveled different paths to the same place.

Someone nurtured those cats in lovingkindness, and lovingkindness is the gift they give, and give again, to their human guardians—that innate peace which has emerged from their inner life in the exchange of love across the species divide. As more and more cats are fortunate enough to be born and to grow in kind and loving care, will their people share with others the lovingkindness that their cats share with them?

Black kitten. White snow. Good luck.

May it be so.

Acknowledgments

My former editor Emily Loose and I thought up the idea for this book together, and we had begun to sketch out a proposal for it when the parent company of the Free Press pulled not just the rug but the whole floor out from under Emily and our whole team there. One day the Free Press existed, and the next day it didn't. I guess that's publishing today.

It took me I don't remember how many drafts to finish a proposal that satisfied my agent, David McCormick, a relentless perfectionist. "More science, Tom," he kept saying, "less of you and that cat." My admiration and affection for David remain nonetheless boundless.

I'm grateful, too, to Mauro DiPreta and Stacy Creamer of Hachette for going for said proposal. Stacy said, "I love the stories about you and Augusta, Tom. Can we have more of those?" Great new editor! Loved cats! We were totally in tune on this book. Then she got a job somewhere else. Publishing today.

After a scary hiatus, during which I begged Mauro please not to hire a cat hater, Michelle Howry arrived, and for the third time—and believe me, luck like this is almost unknown in publishing—I had an editor who simply had everything a writer could want: skill, brains, super-literacy, plus a mysterious way of knowing what I was trying to say when I couldn't quite say it myself.

Now the rest of the Hachette tribe have been emerging, and what an excellent outfit they are. Lauren Hummel can apparently juggle any number of flaming objects. Hallie Patterson is a marketing wiz. Michelle Aielli conducts the orchestra. Mauro DiPreta seems to do everything.

I want to single out two researchers for particular thanks. The findings of Eileen Karsh illuminate the crucial stages of development in very young kittens with a degree of precision that can make an immense difference in the lives of both cats and their owners. Eugenia Natoli's studies of feral cats in Rome, and of the challenges that feral cat colonies present, bring subtlety and wisdom to issues too often oversimplified.

And, as ever, to the best editor of all, my wife, Elizabeth, gratitude beyond expression.

Endnotes

Chapter 1

1. Some disagreement remains about taxonomy, which is something of a scientific swamp. A few authorities insist on designating the domestic cat as a separate species, *Felis catus*. Like most subspecies-vs.-species conundrums, this is really angels dancing on the head of a pin.
2. Hu et al. "Earliest Evidence for Commensal Processes," November 2013.
3. Vigne et al. "Earliest 'Domestic' Cats in China," January 2016.
4. Ibid.
5. Van Neer et al. "More Evidence for Cat Taming," 2014.
6. Blumenstock. "Ten Ways to Unknowingly Crush Your Cat's Spirit," April 2015.
7. Kaplan. "Top 10 Pet-Owner Mistakes," 2011.
8. Grant. "Ten Things Cats Won't Tell You." 2013.
9. Darwin. *Expression of the Emotions,* 2009.
10. Belin et al. "Human Cerebral Response," 2008.
11. Turner and Bateson. *The Domestic Cat,* 1988.
12. Ellis et al. "AAFP and ISFM...Needs Guidelines," 2013.
13. Lue, Pantenburg, and Crawford. "Impact of Owner–Pet Client–Veterinarian Bond," 2008.
14. Bahlig-Pieren and Turner. "Anthropomorphic Interpretations," 1999.
15. Montague et al. "Comparative Analysis of Domestic Cat Genome," December 2014.
16. Bradshaw. *Behaviour of the Domestic Cat,* 1992.

Chapter 2

1. Karsh and Turner. "Human–Cat Relationship," in Turner and Bateson. Unless otherwise specified, all Karsh quotations are from this source.
2. September 19, 1989.

3. Using the Life Satisfaction Index of Neugarten, Havinghurst, and Tobin, 1961.
4. e.g., Mishra with Schroeder. *Measuring the Benefits,* 2014.
5. Turner. "Noncommunicable Diseases," in Zinsstag et al., *One Health,* 2015.
6. McNicholas et al. "Pet Ownership and Human Health," 2005.
7. Koivusilta and Ojanlatva. "To Have or Not to Have a Pet," 2006.
8. Mayon-White. "Pets—Pleasures and Problems," 2005.
9. Clower and Neaves. "Health Care Cost Savings," December 2015.
10. Seymour. "Purr-fect as Pals," 1986.
11. Qureshi et al. "Cat Ownership and Risk," 2009.
12. Thompson. "Most Popular Pet?" 2013.
13. Humane Society of the United States. "Pets by the Numbers," 2016.
14. Voith. "Attachment...to Companion Animals," 1985.
15. http://www.disboards.com/threads/crazy-things-your-cat-does.698221/.

Chapter 3

1. Frazer Sissom, Rice, and Peters. "How Cats Purr," 1991.
2. McComb et al. "Cry Embedded within the Purr," 2009.
3. Budiansky. *If a Lion Could Talk,* 1998.
4. Segelken. "It's the Cat's Meow," 2002.
5. Schötz. "A Phonetic Pilot Study," 2013.
6. Schötz and van de Weijer. "Human Perception of Intonation," 2014.
7. Potter and Mills. "Domestic Cats," 2015.
8. Say, Portier, and Natoli. "High Variation in Multiple Paternity," 1999.
9. Eliot. "In Respect of Felines," 2016.
10. Owren, Rendall, and Ryan. "Redefining Animal Signaling," 2010.
11. Bhalig-Pieren and Turner. "Anthropomorphic Interpretations," 1999.

Chapter 4

1. Driscoll et al. "Near Eastern Origin of Cat Domestication," 2007.
2. Montague et al. "Comparative Analysis," 2014.
3. Driscoll et al. "Taming of the Cat," 2009.
4. Driscoll et al. "From Wild Animals to Domestic Pets," 2009.
5. Montague et al. "Comparative Analysis," 2014.
6. Ewing. "Anti-Anxiety Medication for Cats," 2016.

7. Nagelschneider. *Cat Whisperer,* 2013 (p. 48).
8. Galaxy and Benjamin. *Catification,* 2014 (p. 13).
9. Nagelschneider. *Cat Whisperer* (pp. 61–62).
10. Weir. "Cat Whisperer," 2008.
11. *Cat Daddy* (pp. 55–63).
12. Crank. "Horse Whisperers Part 1," in *Horses and History throughout the Ages,* 2011.
13. Whitaker and Steinkraus. *The Horse,* 2007.
14. (From dust jacket) Brannaman and Reynolds. *Believe: A Horseman's Journey,* 2004.
15. *Cat Daddy* (pp. 144–145).
16. www.bengalcat.co.uk/Bengal-Cat-Breed-Standards.htm.
17. Lipinski et al. "The Ascent of Cat Breeds," 2008.
18. Maldarelli. "Purebred Dogs Can Be Best in Show," 2014.
19. http://cfa.org/Breeders/FAQs/BreedingFAQs.aspx.
20. http://icatcare.org/advice/cat-health/scottish-fold-disease-%E2%80%93-osteochondrodysplasia.
21. http://icatcare.org/advice/cat-health/hypertrophic-cardiomyopathy-hcm-and-testing.
22. http://icatcare.org/advice/cat-health/progressive-retinal-atrophy.
23. http://pets.thenest.com/common-medical-disorders-siamese-cats-7325.html.
24. http://icatcare.org/advice/cat-breeds/inherited-disorders-cats.
25. Levy. "Living-Room Leopards," 2013.
26. http://messybeast.com/twisty.htm.
27. http://messybeast.com/squitten.htm.
28. https://en.wikipedia.org/wiki/Chausie.
29. *Sydney Morning Herald.* "Savannah Cats Banned from Australia," 2008.
30. www.mokavecats.com/.
31. Loxton. *Noble Cat* (p. 51).
32. https://pethelpful.com/exotic-pets/small-exotic-cats.
33. Loxton. *Noble Cat.*
34. Weaver. *personal communication.*
35. http://bigcatrescue.org/hybrid-facts/.
36. Heath. "18 Tigers, 17 Lions,…1 Man Dead in Ohio," 2012.
37. Ibid.
38. Roe. "Keeping a Bobcat or Canadian Lynx as a Pet."
39. www.humanesociety.org/animals/cats/tips/declawing.html.

Chapter 5

1. Natoli et al. "Management of Feral Domestic Cats," 2006.
2. Natoli. "Urban Feral Cats," 1994.
3. Natoli et al. "Management of Feral Domestic Cats," 2006.
4. Cohen. "Caesar's Stabbing Site Identified." 2012.
5. www.romancats.com/torreargentina/en/history.php.
6. www.oecdbetterlifeindex.org/countries/italy/.
7. Silvia Viviani. *personal communication.*
8. www.capitolium.org/eng/fori/cesare.htm.
9. Natoli. "Urban Feral Cats," 1994.
10. Natoli et al. "Relationships between Cat Lovers and Feral Cats," 1999.
11. Turner and Meister. "Hunting Behaviour of Domestic Cat," in Turner and Bateson, 1988.
12. e.g., Liberg and Sandell. "Density, Spatial Organisation, and Reproductive Tactics," in Turner and Bateson, 1988, and Natoli. "Urban Feral Cats," 1994.
13. Clark. *Rome and a Villa.* 1962.
14. Quammen. *Flight of the Iguana.* 1988.
15. http://twenty-somethingtravel.com/2013/04/keats-shelley-and-the-prettiest-cemetery-in-rome/.
16. Hattam. "Extraordinary Lives of Istanbul's Street Cats," 2016.
17. Say, Portier, and Natoli. "High Variation in Multiple Paternity," 1999.
18. Ibid.
19. Levy et al. "Number of Unowned Free-Roaming Cats," 2003.
20. Ramzy. "Australia Defends Plan to Cull Cats." 2015.
21. Toukhatsi, Bennett, and Coleman. "Behaviors, Attitudes toward Semi-owned Cats," 2007.
22. www.dallasnews.com/investigations/problem-solver/20100425-DMN-Problem-Solver-Feral-cat-945.ece.
23. Hildreth, Vantassel, and Hygnstrom. "Feral Cats and their Management," 2010.
24. www.petmd.com/cat/conditions/musculoskeletal/c_ct_hip_dysplasia.
25. Epstein et al. "2015 AAHA/AAFP Pain Management Guidelines," 2015.
26. http://dallasanimalservices.org/.
27. www.lifelineanimal.org/outreach/catlanta.
28. www.baltimorecats.org/stray.htm.

29. www.treehouseanimals.org/site/PageServer?pagename=programs _trap_neuter_return_feral_cats.
30. www.feralcatcaretakers.org/Caretaking/InterestedParties.html.
31. www.humanesociety.org/issues/pet_overpopulation/facts/pet _ownership_statistics.html.
32. www.aspca.org/animal-homelessness/shelter-intake-and-surrender/pet-statistics.
33. www.pbs.org/newshour/bb/why-activists-are-fighting-over-feral -felines-2/.
34. Barcott. "Kill the Cat that Kills the Bird?" 2007.
35. Manning. "Kittens Dropped at Shelters," 2014.
36. Loss, Will, and Marra. "Impact of Free-Ranging Domestic Cats," 2013.
37. Angier. "Cuddly Kitty Is Deadlier than You Think," 2013.
38. Loss, Will, and Marra. "Estimation of Bird–Vehicle Collision Mortality," 2014.
39. Loss et al. "Bird–Building Collisions," 2014.
40. https://wwwrspb.org.uk/get-involved/community-and-advice/ garden-advice/unwantedvisitors/cats/birddeclines.aspx.
41. Nogales et al. "Feral Cats and Biodiversity Conservation," 2013.
42. Ibid.
43. Beckerman, Boots, and Gaston. "Urban Bird Declines and Fear of Cats," 2007.
44. Gerhold and Jessup. "Zoonotic Diseases," 2012.
45. Sugden et al. "Is Toxoplasma Gondii Infection Related to Brain and Behavior Impairments in Humans?" 2016.
46. Nutter, Levine, and Stoskopf. "Reproductive Capacity," 2004.
47. PETA. "Feral Cats: Trapping Is Kindest Solution."
48. PETA. "Animal Rights Uncompromised."
49. Natoli et al. "Management of Feral Domestic Cats," 2006.
50. Chu, Anderson, and Rieser. "Population Characteristics and Neuter Status," 2009.
51. Patronek. "Mapping and Measuring Disparities," 2010.
52. Levy, Gale, and Gale. "Evaluation of the Effect of a Long-Term Trap-Neuter-Return," 2003.
53. www.chemnet.com/Global/Offer-to-Sell/Benzyldiethyl%28%282,6-x ylylcarbamoyl%29-methyl%29-ammonium-benzoate-2369625.html.
54. Hildreth, Vantassel, and Hygnstrom. "Feral Cats and Their Management," 2010.

55. Peterson et al. "Opinions from the Front Lines," 2012.
56. American Veterinary Medical Association. "Free-Roaming and Abandoned Cats," 2016.
57. Mooney. "The Science of Why We Don't Believe Science," 2011.
58. PETA. "Animal Rights Uncompromised: Feral Cats."
59. PETA. "Feral Cats: Trapping Is Kindest Solution."
60. Willson, Okunlola, and Novak. "Birds Be Safe," 2015.
61. Kays et al. "Cats Are Rare Where Coyotes Roam," 2015.
62. Jonathan Young, Presidio Trust biologist, *personal communication*.
63. http://www.alleycat.org/resources/feral-cat-activist-winter-2008/.
64. http://www.alleycat.org/resources/cat-licensing-a-license-to-kill/.
65. www.petfinder.com/dogs/lost-and-found-dogs/microchip-faqs/.
66. http://pets.costhelper.com/spay-neuter-cat.html.

Chapter 6

1. Loyd et al. "Risk Behaviours," 2013.
2. *Montana Pioneer.* "Livingston Man Sentenced," 2007.
3. www.livingstonenterprise.com/content/log-book-%E2%80%94-july-22-27.
4. Bilefsky. "London's Cats Falling Victim," 2016.
5. Loyd et al. "Quantifying Free-roaming Domestic Cat Predation," 2013.
6. Horn et al. "Home Range, Habitat Use, Activity Patterns," 2011.
7. McDonald et al. "Reconciling Actual and Perceived Rates," 2015.
8. Tschanz et al. "Hunters and Non-hunters," 2011.
9. Loss, Will, and Marra, "Impact of Free-ranging Domestic Cats," 2013.
10. American Bird Conservancy. "Human Attitudes and Behavior Regarding Cats," 1997.
11. Patronek, Beck, and Glickman. "Dynamics of Dog and Cat Populations," 1997.
12. Hall et al. "Community Attitudes and Practices," 2016.
13. PetMD.com. "How to Walk a Cat (and Live to Tell about It)." www.petmd.com/cat/training/evr_ct_how_to_walk_you_walk_your_cat?page=show.
14. Lord et al. "Search and Identification Methods," 2007.
15. Belluck. "A Cat's 200-Mile Trek," 2013.
16. Holland. "Watch," 2014.

17. Much guidance comes from web page: Miller. "How to Find a Lost Cat." *MyPetMD.com.*
18. Lord et al. "Search and Identification Methods," 2007.
19. Belluck. "Cat's 200-Mile Trek," 2013.
20. Adapted from a variety of sources; primary one is Ellis et al. "AAFP and ISFM...Needs Guidelines," 2013.
21. Many examples from Herron and Buffington. "Environmental Enrichment for Indoor Cats," 2010.
22. Morris. "Idiosyncratic Nutrient Requirements," 2002.
23. Bradshaw. *Behaviour of the Domestic Cat,* 1992.
24. Hewson-Hughes et al. "Balancing Macronutrient Intake," 2016.
25. Association for Pet Obesity Prevention. "An Estimated 58% of Cats and 54% of Dogs."
26. Pryor. *Getting Started: Clicker Training for Cats,* 2003.
27. Tellington-Jones with Taylor. *The Tellington TTtouch* [sic]: *Caring for Animals,* 1992.
28. Tellington-Jones. *The Tellington TTouch* [sic] *for Happier, Healthier Cats.*
29. Ellis et al. "AAFP and ISFM...Needs Guidelines," 2013.
30. Budiansky. *The Character of Cats,* 2002.
31. Bradshaw. *Cat Sense,* 2013.
32. ASPCA.org. Shelter intake and surrender.
33. Abbe. "Walker Art Center," 2016.
34. https://en.wikipedia.org/wiki/Cats_and_the_Internet.
35. http://tubularinsights.com/2-million-cat-videos-youtube/.

Chapter 7

1. Bellows et al. "Evaluating Aging in Cats: How to Determine What Is Healthy and What Is Disease," 2016.
2. Merola and Mills. "Behavioural Signs of Pain," 2016.
3. Coates. "How to Know When a Cat Is Hurting." 2016.
4. Marek. "Experts Find New Ways to Assess Pain," 2016.
5. Epstein, et al. "2015 AAHA/AAFP Pain Management Guidelines." 2015.
6. Singer. "How to Recognize an Emergency," 2016.
7. Merola and Mills. "Behavioural Signs of Pain," 2016.
8. Lindner. "Creating the Trauma Initiative," 2016
9. Blumenstock. "Senior Moments," 2016.

10. Remitz. "Defining Senior Age in Cats," 2016.
11. Intile. "How the 'Will Rogers Phenomenon,'" 2016.
12. Vogelsang. "'Miracle' Technology," 2016.
13. Villalobos. "Quality of Life Scale" from *Canine and Feline Geriatric Oncology,* 2007. Revised 2011.
14. Leary et al. *AVMA Guidelines for the Euthanasia,* 2013.
15. Stewart. "Loss of a Pet—Loss of a Person" in Katcher and Beck, 1983.
16. Ibid.
17. Field et al. "Role of Attachment in Response to Pet Loss," 2009.
18. Quackenbush and Glickman. "Social Work Services," in Katcher and Beck, 1983.

Chapter 8

1. www.humanesociety.org/issues/pet_overpopulation/facts/pet _ownership_statistics.html.
2. www.aspca.org/animal-homelessness/shelter-intake-and-surrender/ pet-statistics.
3. www.mapsofworld.com/world-top-ten/countries-with-most-pet -cat-population.html.
4. Katcher. "Man and the Living Environment," in Katcher and Beck, 1983. All following Katcher quotations are from same.
5. Voith. "Attachment of People to Companion Animals," 1985.
6. http://islam.ru/en/content/story/love-and-importance-cats-islam, http://www.muslimheritage.com/article/cats-islamic-culture.
7. Fisher. *Why We Love,* 2004.
8. Zak. "Dogs (and Cats) Can Love," 2014.
9. Levinson. "The Future of Research into Relationships," 1982.
10. *New York Times,* April 3, 1984.
11. Knight. Comments on Katcher's "Excursion into Cyclical Time," in Katcher and Beck, 1983.
12. Wiman. *My Bright Abyss,* 2013.
13. Brody. "Coloring Your Way through Grief," 2016.
14. www.behindthename.com/name/isabel.

Bibliography

Abbe, Mary. "Goodbye, Kitty: Walker Art Center is Ending Internet Cat Video Festival." Minneapolis *Star Tribune,* March 16, 2016.

Alley Cat Allies. "Case Study of a Feral Cat Sanctuary." http://www.alleycat.org/resources/feral-cat-activist-winter-2008/, February 2008.

Alley Cat Allies. "Cat Licensing: A License to Kill." http://www.alleycat.org/resources/cat-licensing-a-license-to-kill/. N.D.

Alley Cat Allies. "The University of Nebraska is Dangerously Wrong about Feral Cats." *Alleycat.org,* 2010.

American Bird Conservancy. "Human Attitudes and Behavior Regarding Cats." 1997. Republished at www.njaudubon.org/portals/10/catsindoors/pdf/attitude.pdf.

American Bird Conservancy. "Letter to Secretary of the Interior Sally Jewell, Urging Swift Action to Address the Threat to Wildlife Populations and Human Health Posed by Feral Cats." January 28, 2014.

American Veterinary Medical Association. "Free-Roaming and Abandoned Cats." Policy Statement. February 10, 2016.

American Veterinary Medical Association. "Raw or Undercooked Animal-Source Protein in Cat and Dog Diets." August 2012.

Angier, Natalie. "That Cuddly Kitty Is Deadlier than You Think." *New York Times,* January 29, 2013.

Animal Welfare Institute. "Cats" in *Comfortable Quarters for Laboratory Animals.* Washington, DC: Animal Welfare Institute, 1979.

Aschwanden, Christie. "Science Isn't Broken, It's Just a Hell of a Lot Harder than We Give It Credit for." *FiveThirtyEight.com,* August 19, 2015.

Ash, Sara J., and Clark E. Adams. "Public Preferences for Free-Ranging Domestic Cat (*Felis catus*) Management Options." *Wildlife Society Bulletin* 31, 2003.

ASPCA.org. "Shelter Intake and Surrender."

Associated Press. "Feral Cats Should Be Killed with 'a Gunshot to the Head' to Control Population: UNL Undergraduates Report." *Huffington Post,* December 2, 2010.

Association for Pet Obesity Prevention. "An Estimated 58% of Cats and 54% of Dogs in the United States are Overweight or Obese." *Petobesityprevention.org.*

Axelrod, Julie. "Grieving the Loss of a Pet." *PsychCentral.com,* October 2015.

Bahlig-Pieren, Zana, and Dennis C. Turner. "Anthropomorphic Interpretations and Ethological Descriptions of Dog and Cat Behaviour by Lay People." *Anthrozoös* 12, 1999.

Barcott, Bruce. "Kill the Cat that Kills the Bird?" *New York Times,* December 2, 2007.

Barratt, David G. "Home Range Size, Habitat Utilisation, and Movement Patterns of Suburban and Farm Cats *Felis catus.*" *Ecography* 20, 1997.

Bateson, P., M. Mendl, and J. Feaver. "Play in the Domestic Cat Is Enhanced by Rationing of Mother during Lactation." *Animal Behaviour* 40, 1990.

Beadle, Muriel. *The Cat: History, Biology, and Behavior.* New York: Simon and Schuster, 1977.

Beaver, Bonnie V. *Feline Behavior: A Guide for Veterinarians.* Philadelphia: W. B. Saunders, 1992.

Beckerman, A. P., M. Boots, and K. J. Gaston. "Urban Bird Declines and the Fear of Cats." *Animal Conservation* 10, 2007.

Behrend, K., and M. Wegler. *Complete Book of Cat Care: How to Raise a Happy and Healthy Cat.* Hauppauge, NY: Barron's Educational Series, Inc., 1991.

Bekoff, Marc. *The Emotional Lives of Animals.* Novato, CA: New World Library, 2007.

Belin, Pascal, et al. "Human Cerebral Response to Animal Affective Vocalizations." *Proceedings of the Royal Society B* 275, 2008.

Bellows, Jan, et al. "Evaluating Aging in Cats: How to Determine What Is Healthy and What Is Disease." *Journal of Feline Medicine and Surgery,* July 2016.

Belluck, Pam. "A Cat's 200-Mile Trek Home Leaves Scientists Guessing." *New York Times,* January 19, 2013.

Bergstrom, Dana M., et al. "Indirect Effects of Invasive Species Removal Devastate World Heritage Island." *Journal of Applied Ecology* 46, 2009.

Bilefsky, Dan. "London's Cats Are Falling Victim to a Two-Legged Predator." *New York Times,* May 13, 2016.

Blackshaw, J. K. "Abnormal Behavior in Cats." *Australian Veterinary Journal* 65, 1988.

Blumenstock, Kathy. "Senior Moments: Keeping Your Older Cat Young at Heart." *Pet360.com,* March 8, 2016.

Blumenstock, Kathy. "Ten Ways to Unknowingly Crush Your Cat's Spirit." *Pet360.com,* April 20, 2015.

Bonanni, Roberto, et al. "Feeding Order in an Urban Feral Domestic Cat Colony: Relationship to Dominance Rank, Sex, and Age." *Animal Behaviour* 74, 2007.

Borchelt, Peter L., and Victoria L. Voith. "Aggressive Behavior in Cats" in *Readings in Companion Animal Behavior.* Trenton, NJ: Veterinary Learning Systems, 1996.

Borchelt, Peter L., and Victoria L. Voith. "Elimination Behavior Problems in Cats." In *Readings in Companion Animal Behavior.* Trenton, NJ: Veterinary Learning Systems, 1996.

Bradshaw, John W. S. *Cat Sense: How the New Feline Science Can Make You a Better Friend to Your Pet.* New York: Basic Books, 2013.

Bradshaw, John W. S. *The Behaviour of the Domestic Cat.* Wallingford, UK: CAB International, 1992.

Brannaman, Buck, and William Reynolds. *Believe: A Horseman's Journey.* Guilford, CT: The Lyons Press, 2004.

Broad, Michael. "Some Indoor/Outdoor Cat Facts." *Pictures-of-cats .org,* July 30, 2012.

Brody, Jane E. "Coloring Your Way through Grief." *New York Times,* May 17, 2016.

Bryan, Kathleen J. "Butte Cat Hoarder Gets Suspended Sentence for 'Severe' Neglect, Abuse of Felines." *Montana Standard,* May 13, 2016.

Budiansky, Stephen. *If a Lion Could Talk: Animal Intelligence and the Evolution of Consciousness.* New York: Free Press, 1998.

Budiansky, Stephen. *The Character of Cats: The Origins, Intelligence, Behavior, and Stratagems of Felis silvestris catus.* New York: Viking, 2002.

Budiansky, Stephen. *The Nature of Horses: Exploring Equine Evolution, Intelligence, and Behavior.* New York: Free Press, 1997.

Budiansky, Stephen. *The World According to Horses: How They Run, See, and Think.* New York: Henry Holt, 2000.

Burns, Katie. "AVMA Revises Policy on Feral Cats to Encourage Collaboration." *Journal of the American Veterinary Medical Association*, March 1, 2016.

Cafazzo, S., and E. Natoli. "The Social Function of Tail Up in the Domestic Cat." *Behavioural Processes* 80, 2009.

Calver, Michael, et al. "Reducing the Rate of Predation on Wildlife by Pet Cats: The Efficacy and Practicability of Collar-Mounted Pounce Protectors." *Biological Conservation* 137, 2007.

Carlstead, K., J. L. Brown, and W. Strawn. "Behavioral and Physiological Correlates of Stress in Laboratory Cats." *Applied Animal Behaviour Science* 38, 1993.

Caro, T. M., and M. D. Hauser. "Is There Teaching in Nonhuman Animals?" *The Quarterly Review of Biology* 67, 1992.

Caro, T. M. "Predatory Behaviour and Social Play in Kittens." *Behaviour* 76, 1981.

Centonze, L. A., and J. K. Levy. "Characteristics of Free-Roaming Cats and Their Caretakers." *Journal of the American Veterinary Medical Association*, June 2002.

Chaban, Matt A. V. "Gentrification's Latest Victims: New York's Feral Cats." *New York Times*, May 24, 2016.

Cheney, Carolyn M., et al. "A Large Granular Lymphoma and Its Derived Cell Line." *In Vitro Cellular & Developmental Biology* 26, 1990.

Chu, Karyen, Wendy M. Anderson, and Micha Y. Rieser. "Population Characteristics and Neuter Status of Cats Living in Households in the United States." *Journal of the American Veterinary Medical Association* 234, 2009.

Clark, Eleanor. *Rome and a Villa*. New York: Atheneum, 1962.

Clower, Terry L., and Tonya T. Neaves. "The Health Care Cost Savings of Pet Ownership." *Human Animal Bond Research Initiative Foundation*, December 2015.

Coates, Jennifer. "Defining an Adoptable Animal." *PetMD.com*, January 17, 2016.

Coates, Jennifer. "How to Know When a Cat is Hurting." *PetMD.com*, April 11, 2016.

Coates, Jennifer. "Veterinary Hospice Care Is Beautiful When Done Well." *PetMD.com*, March 21, 2016.

Cohen, Jennie. "Julius Caesar's Stabbing Site Identified." *History.com*, October 11, 2012.

Coleman, John S., and Stanley A. Temple. "Rural Residents' Free-Ranging Domestic Cats: A Survey." *Wildlife Society Bulletin* 21, 1993.

Colleran, Elizabeth. "Feline Posture: A Visual Dictionary." *Dvm360 .com*, 2015.

Collier, Glen E., and Stephen J. O'Brien. "A Molecular Phylogeny of the Felidae: Immunological Distance." *Evolution* 39, 1985.

Cornell University College of Veterinary Medicine. "Cats and Cucumbers—So What's the Big Deal?" *Vet.cornell.edu*, December 7, 2015.

Cornell University College of Veterinary Medicine Feline Health Center. "Is It Dementia or Normal Aging?" *CatWatch*, May 2016.

Cornell University College of Veterinary Medicine Feline Health Center. "Shelter Alternatives Can Save Lives." *CatWatch*, February 2016.

Courchamp, Franck, Michel Langlais, and George Sugihara. "Cats Protecting Birds: Modelling the Mesopredator Release Effect." *Journal of Animal Ecology* 68, 1999.

Cox, Ana Marie, Interviewer. "Jackson Galaxy Thinks Cats Saved His Life." *New York Times Magazine*, December 17, 2015.

Crank, Cindy. "Horse Whisperers Part 1: Origins, Societies, and Secrets," in *Horses and History throughout the Ages, Horsesandhistory .wordpress.com*, April 5, 2011.

Crooks, Kevin R., and Michael E. Soulé. "Mesopredator Release and Avifaunal Extinctions in a Fragmented Ecosystem." *Nature* 400, 1999.

Darnton, Robert. *The Great Cat Massacre: And Other Episodes in French Cultural History.* New York: Basic Books, 1985.

Darwin, Charles. *The Expression of the Emotions in Man and Animals.* New York: Penguin Classics, 2009.

DeLuca, A. M., and K. C. Kranda. "Environmental Enrichment in a Large Animal Facility." *Lab Animal* 21, 1992.

Disboards.com. "Crazy Things Your Cat Does." http://www.disboards. com/threads/crazy-things-your-cat-does.698221/.

Dore, F. Y. "Search Behavior of Cats (*Felis catus*) in an Invisible Displacement Test: Cognition and Experience." *Canadian Journal of Psychology* 44, 1990.

Driscoll, Carlos A., et al. "From Wild Animals to Domestic Pets, an Evolutionary View of Domestication." *Proceedings of the National Academy of Sciences of the United States of America*, 2009.

Bibliography

Driscoll, Carlos A., et al. "The Near Eastern Origin of Cat Domestication." *Science,* July 27, 2007.

Driscoll, Carlos A., et al. "The Taming of the Cat." *Scientific American,* June 2009.

Drouard, C. M. "Behavior of Housed and Stray Cats." ("Le Comportement du Chat et les Chats Errants.") Thesis. Alfort, France: École Nationale Vétérinaire, 1979.

Eliot, T. S. "In Respect of Felines." *New York Review of Books,* January 14, 2016.

Eliot, T. S. *Old Possum's Book of Practical Cats.* London: Faber & Faber, 1939.

Ellis, Sarah L. H., et al. "AAFP (American Association of Feline Practitioners) and ISFM (International Society of Feline Medicine) Feline Environmental Needs Guidelines." *Journal of Feline Medicine and Surgery* 15, 2013.

Epstein, Mark E., et al. "2015 AAHA/AAFP Pain Management Guidelines for Dogs and Cats." *Journal of Feline Medicine and Surgery* 17, 2015.

Ewing, Tom. "Anti-Anxiety Medication for Cats." *Catnip* (newsletter of the Cummings School of Veterinary Medicine at Tufts University), February 24, 2016.

Examiner.com. "University of Nebraska Report Recommends Killing Feral Cats for Control." December 2, 2010.

Farm Animal Welfare Council. *Farm Animal Welfare in Great Britain: Past, Present, and Future.* London: Farm Animal Welfare Council, 2009.

Ferrari, Pier Francesco, and Giacomo Rizzolatti. "Mirror Neuron Research: The Past and the Future." *Philosophical Transactions of the Royal Society B: Biological Sciences* 369, 2014.

Field, Nigel P., et al. "Role of Attachment in Response to Pet Loss." *Death Studies* 33, 2009.

Fisher, Helen. *Why We Love: The Nature and Chemistry of Romantic Love.* New York: Henry Holt and Co., 2004.

Foderaro, Lisa W. "At a Long Island Beach, Human Tempers Flare over Claws and Feathers." *New York Times,* April 18, 2015.

Fogle, Bruce. *The Cat's Mind: Understanding Your Cat's Behavior.* New York: Howell Book House, 1992.

Fogle, Bruce. *The Complete Illustrated Guide to Cat Care and Behavior.* San Diego: Thunder Bay Press, 1999.

Foster, Derek, et al. "'I Can Haz Emoshuns?'—Understanding Anthropomorphosis of Cats among Internet Users." *IEEE* (Institute of Electrical and Electronics Engineers Computer Society) *International Conference on Privacy, Security, Risk, and Trust,* and *IEEE International Conference on Social Computing.* 2011.

Fox, M. W. "The Behavior of Cats." In E. S. E. Hafez, ed., *The Behavior of Domestic Animals,* 3rd ed. London: Bailliere Tindall, 1975.

Frazer Sissom, Dawn E., D. A. Rice, and G. Peters. "How Cats Purr." *Journal of Zoology* 223, 1991.

Frazier, A., and N. Eckroate. "Desirable Behavior in Cats and Owners." In Frazier, Anitra, and Eckroate, *The New Natural Cat: A Complete Guide for Finicky Owners.* New York: Penguin Books, 1990.

Freedman, Adam H., et al. "Genome Sequencing Highlights the Dynamic Early History of Dogs." *PLOS Genetics* 10, 2014.

Galaxy, Jackson, and Kate Benjamin. *Catification: Designing a Happy and Stylish Home for Your Cat (and You!).* New York: Jeremy P. Tarcher / Penguin, 2014.

Galaxy, Jackson, with Joel Derfner. *Cat Daddy: What the World's Most Incorrigible Cat Taught Me about Life, Love, and Coming Clean.* New York: Jeremy P. Tarcher / Penguin, 2012.

Gerhold, R. W., and D. A. Jessup. "Zoonotic Diseases Associated with Free-Roaming Cats." *Zoonoses and Public Health.* Blackwell Verlag GmbH, 2012.

Gehrt, Stanley D., et al. "Population Ecology of Free-Roaming Cats and Interference Competition by Coyotes in Urban Parks." *PLOS ONE* 8, 2013.

Glass, Gregory E., et al. "Trophic Garnishes: Cat–Rat Interactions in an Urban Environment." *PLOS ONE* 4, 2009.

Grant, Kelli B. "Ten Things Cats Won't Tell You." *Marketwatch.com.* 2013.

Grubbs, Shannon E., and Paul R. Krausman. "Observations of Coyote–Cat Interactions." *Journal of Wildlife Management* 73, 2009.

Guyot, G. W., T. L. Bennett, and H. A. Cross. "The Effects of Social Isolation on the Behavior of Juvenile Domestic Cats." *Developmental Psychobiology* 13, 1980.

Hall, Catherine M., et al. "Community Attitudes and Practices of Urban Residents Regarding Predation by Pet Cats on Wildlife: An International Comparison." *PLOS ONE* 10, 2016.

Hanna, Jack, with Amy Parker. *Jungle Jack: My Wild Life.* Nashville: Thomas Nelson, 2008.

Hart, B. L. *Feline Behavior: Collected Columns from Feline Practice Journal*. Santa Barbara, CA: Veterinary Practice Publishing Co., 1978.

Hartwell, Sarah. "Cat Communication." *Messybeast.com*, 1995–2012.

Hartwell, Sarah. "Do Cats Have Emotions?" *Messybeast.com*, 2001–2013.

Hattam, Jennifer. "The Extraordinary Lives of Istanbul's Street Cats." *Citylab.com*, April 26, 2016.

Hauser, Marc D. *Wild Minds: What Animals Really Think*. New York: Henry Holt, 2000.

Hayes, A. A. "Keeping Your Cat Healthy: Play/Exercise." In Kay, W. J., and E. Randolph, eds., *The Complete Book of Cat Health*. New York: Macmillan, 1985.

Heath, Chris. "18 Tigers, 17 Lions, 8 Bears, 3 Cougars, 2 Wolves, 1 Baboon, 1 Macaque, and 1 Man Dead in Ohio." *GQ*, March 2012.

Heffner, Henry E., and Rickye S. Heffner. "The Behavioral Study of Mammalian Hearing." In Popper, Arthur N., and Richard R. Fay, eds. *Perspectives on Auditory Research*. New York: Springer Science + Business Media, 2014.

Heffner, Rickye S., and Henry E. Heffner. "Hearing in Mammals: The Least Weasel." *Journal of Mammalogy* 66, 1985.

Heffner, Rickye S., and Henry E. Heffner. "Hearing Range of the Domestic Cat." *Hearing Research* 19, 1985.

Heffner, Rickye S., and Henry E. Heffner. "Sound Localization Acuity in the Cat: Effect of Azimuth, Signal Duration, and Test Procedure." *Hearing Research* 36, 1988.

Heffner, Rickye S., and Henry E. Heffner. "The Sound-Localization Ability of Cats." Letter to the Editor. *Journal of Neurophysiology* 94, 2005.

Heffner, Rickye S., and Henry E. Heffner. "Visual Factors in Sound Localization in Mammals." *Journal of Comparative Neurology* 317, 1992.

Hendricks, Cleon G., et al. "Tail Vaccination in Cats: A Pilot Study." *Journal of Feline Medicine and Surgery*. Published online before print, October 2013.

Herrick, Francis H. "Homing Powers of the Cat." *The Scientific Monthly*, June 1922.

Herron, Meghan E., and C. A. Tony Buffington. "Environmental Enrichment for Indoor Cats." *Compendium: Continuing Education for Veterinarians*. American Association of Feline Practitioners, December 2010.

Heuer, Victoria. "Crate Traveling for Cats." *PetMD.com,* July 19, 2016.

Heuer, Victoria. "Ten Things to Consider before Bringing a New Pet Home." *PetMD.com,* April 12, 2016.

Hewson-Hughes, Adrian K., et al. "Balancing Macronutrient Intake in a Mammalian Carnivore: Disentangling the Influences of Flavour and Nutrition." *Royal Society Open Science,* June 15, 2016.

Hildreth, Aaron M., Stephen M. Vantassel, and Scott E. Hygnstrom. "Feral Cats and Their Management." University of Nebraska, Lincoln Extension, 2010.

Holland, Jennifer S. "Watch: How Far Do Your Cats Roam? The New Cat Tracker Project Maps Outdoor Movements of Pet Felines." *NationalGeographic.com News,* August 8, 2014.

Horn, Jeff A., et al. "Home Range, Habitat Use, and Activity Patterns of Free-Roaming Domestic Cats." *Journal of Wildlife Management* 75, 2011.

Horowitz, Alexandra. *Inside of a Dog: What Dogs See, Smell, and Know.* New York: Scribner, 2009.

Houser, Susan. "Counting Feral Cats." *Huffington Post,* February 29, 2016.

Hu, Yaowu, et al. "Earliest Evidence for Commensal Processes of Cat Domestication." *Proceedings of the National Academy of Sciences,* Early Edition, November 13, 2013.

Hughes, Nelika K., Catherine J. Price, and Peter B. Banks. "Predators Are Attracted to the Olfactory Signals of Prey." *PLOS ONE* 5, 2010.

Humane Society of the United States. "How to Keep Your Cat Happy Indoors." *Humanesociety.org,* July 2013.

Humane Society of the United States. "Pets by the Numbers: U.S. Pet Ownership, Community Cat and Shelter Population Estimates." 2016.

Indiana University. "Not-So-Guilty Pleasure: Viewing Cat Videos Boosts Energy and Positive Emotions, IU Study Finds." Press Release, June 18, 2015.

Intile, Joanne. "How the 'Will Rogers Phenomenon' Affects Your Pet's Cancer Diagnosis." *PetMD.com,* March 30, 2016.

Ioannidis, John P. A. "Why Most Published Research Findings are False." *PLOS Medicine* 2, August 2005.

Jackson, Peter, and Kristin Nowell. "Wild Cats: Status Survey and Conservation Action Plan." *IUCN/SSC Action Plans for the Conservation of Biological Diversity.* June 28, 1996.

Johns, Catherine. "The Tombstone of Laetus' Daughter: Cats in Gallo-Roman Sculpture." *Britannia* 34, 2003.

Kaplan, Megan. "The Top 10 Pet-Owner Mistakes." *Articles.cnn.com,* 2011.

Karagiannis, Christos, and Daniel Mills. "Feline Cognitive Dysfunction Syndrome." *Veterinary Focus* 24, 2014.

Karsh, E. B. "The Effects of Early Handling on the Development of Social Bonds between Cats and People." In Katcher and Beck, *New Perspectives on Our Lives with Companion Animals.* Philadelphia: University of Pennsylvania Press, 1983.

Karsh, E. "Factors Influencing the Socialization of Cats to People." In Anderson, R. K., B. L. Hart, and L. A. Hart, eds. *The Pet Connection: Its Influence on Our Health and Quality of Life.* Minneapolis: University of Minnesota Press, 1984.

Karsh, Eileen B., and Dennis C. Turner. "The Human–Cat Relationship." In Turner and Bateson, *The Domestic Cat: The Biology of Its Behaviour.* Cambridge, UK: Cambridge University Press, 1988.

Katcher, Aaron Honori. "Man and the Living Environment: An Excursion into Cyclical Time." In Katcher and Beck, 1983.

Katcher, Aaron Honori, and Alan M. Beck, eds., *New Perspectives on Our Lives with Companion Animals.* Philadelphia: University of Pennsylvania Press, 1983.

Kays, Roland, et al. "Cats are Rare where Coyotes Roam." *Journal of Mammalogy* 96, June 2015.

Khuly, Patty. "Should We Let Pets Have Sex with Each Other?" *PetMD .com,* January 5, 2016.

King, Barbara J. "When Animals Mourn." *Scientific American,* July 2013.

Kitchener, Andrew. *The Natural History of the Wild Cats.* Ithaca, NY: Comstock Publishing Associates, 1991.

Knight, James A. "Comments on Aaron Katcher's 'Excursion into Cyclical Time.'" In Katcher and Beck, 1983.

Koivusilta, Leena K., and Ansa Ojanlatva. "To Have or Not to Have a Pet for Better Health?" *PLOS ONE* 1, 2006.

Kramer, David F. "Is It Safe for Your Cat to Eat Bugs?" *PetMD.com,* August 15, 2016.

Krieger, Marilyn. "Why Do Cats Purr?" *Catster,* March 21, 2014.

Kristensen, F. and J. A. Barsanti. "Analysis of Serum Proteins in Clinically Normal Pet and Colony Cats, using Agarose Electrophoresis." *American Journal of Veterinary Research* 38, 1977.

Kubota, K. "Physiology of the Brain: Intelligent Behavior and Brain." *Nippon Rinsho* 45, 1987.

Kupper, W. "Keeping Laboratory Cats under the Right Conditions." (Zur artgemassen und verhaltensgerechten Unterbringung von Versuchskatzen.) *Du und das Tier* 9, 1979.

Lauber, T. Bruce, and Barbara A. Knuth. "The Role of Ethical Judgments Related to Wildlife Fertility Control." *Society and Natural Resources* 20, 2007.

Lazenby, Billie T., Nicholas J. Mooney, and Christopher R. Dickman. Effects of Low-Level Culling of Feral Cats in Open Populations: A Case Study from the Forests of Southern Tasmania." *Wildlife Research* 41, 2014.

Leary, Steven, et al. *AVMA Guidelines for the Euthanasia of Animals: 2013 Edition.* Schaumburg, IL: American Veterinary Medical Association, 2013.

Legambiente. *IV Rapporto Nazionale: Animali in Città.* March 3, 2015.

Lei, Weiwei. "Functional Analyses of Bitter Taste Receptors in Domestic Cats (*Felis catus*)." *PLOS ONE* 10, 2015.

Levine, Emily D., et al. "Owner's Perception of Changes in Behaviors Associated with Dieting in Fat Cats." *Journal of Veterinary Behavior* 11, 2016.

Levinson, Boris M. "The Future of Research into Relationships between People and their Animal Companions." *International Journal for the Study of Animal Problems* 3, 1982.

Levy, Ariel. "Living-Room Leopards." *The New Yorker,* May 6, 2013.

Levy, Julie K., David W. Gale, and Leslie A. Gale. "Evaluation of the Effect of a Long-Term Trap-Neuter-Return and Adoption Program on a Free-Roaming Cat Population." *Journal of the American Veterinary Medical Association* 222, 2003.

Levy, Julie K., et al. "Number of Unowned Free-Roaming Cats in a College Community in the Southern United States and Characteristics of Community Residents Who Feed Them." *Journal of the American Veterinary Medical Association* 223, 2003.

Liberg, Olof, and Mikael Sandell. "Density, Spatial Organisation, and Reproductive Tactics in the Domestic Cat and Other Felids." In Turner and Bateson, 1988.

Lindner, Larry. "Creating the Trauma Initiative." *Catnip.* April 24, 2016.

Lipinski, Monika J., et al. "The Ascent of Cat Breeds: Genetic Evaluations of Breeds and Worldwide Random Bred Populations." *Genomics* 91, 2008.

Lock, Cheryl. "How to Socialize a Kitten with New People." *PetMD* *.com,* October 8, 2015.

Lohr, Cheryl A., Linda J. Cox, and Christopher Al Lepczyk. "Costs and Benefits of Trap-Neuter-Release and Euthanasia for Removal of Urban Cats in Oahu, Hawaii." *Conservation Biology* 27, 2012.

Longcore, Travis, Catherine Rich, and Lauren M. Sullivan. "Critical Assessment of Claims Regarding Management of Feral Cats by Trap-Neuter-Return." *Conservation Biology* 23, 2009.

Lord, Linda K., et al. "Search and Identification Methods that Owners Use to Find a Lost Cat." *Journal of the American Veterinary Medical Association* 230, no. 2, 2007.

Loss, Scott R., et al. "Bird–Building Collisions in the United States: Estimates of Annual Mortality and Species Vulnerability." *The Condor* 116, 2014.

Loss, Scott R., Tom Will, and Peter P. Marra. "Estimation of Bird–Vehicle Collision Mortality on U.S. Roads." *Journal of Wildlife Management* 78, 2014.

Loss, Scott R., Tom Will, and Peter P. Marra. "The Impact of Free-Ranging Domestic Cats on Wildlife of the United States." *Nature Communications* 4, 2013.

Loveridge, G., L. J. Horrocks, and A. J. Hawthorne. "Environmentally Enriched Housing for Cats When Housed Singly." *Animal Welfare* 4, 1995.

Loveridge, G. "Provision of Environmentally Enriched Housing for Cats." *Animal Technology: Journal of the Institute of Animal Technology* 45, 1994.

Loxton, Howard. *The Noble Cat.* New York: Portland House, 1990.

Loyd, K. A. T., and J. L. DeVore. "An Evaluation of Feral Cat Management Options Using a Decision Analysis Network." *Ecology and Society* 15, 2010.

Loyd, K. A. T., and S. M. Hernandez. "Public Perceptions of Domestic Cats and Preferences for Feral Cat Management in the Southeastern United States." *Anthrozoös* 25, 2012.

Loyd, K. A. T., et al. "Quantifying Free-Roaming Domestic Cat Predation Using Animal-Borne Video Cameras." *Biological Conservation* 160, 2013.

Loyd, K. A. T., et al. "Risk Behaviours Exhibited by Free-Roaming Cats in a Suburban U.S. Town." *Veterinary Record* 10, 2013.

Lue, Todd W., Debbie P. Pantenburg, and Phillip M. Crawford. "Impact of the Owner–Pet and Client–Veterinarian Bond on the Care that Pets Receive." *Journal of the American Veterinary Medical Association* 232, "Vet Med Today: Special Report," 2008.

Luescher, U. A., D. B. McKeown, and J. Halip. "Stereotypic or Obsessive-Compulsive Disorders in Dogs and Cats." *Veterinary Clinics of North America: Small Animal Practice* 21, 1991.

Macdonald, David, and Andrew Loveridge. *The Biology and Conservation of Wild Felids.* New York: Oxford University Press, 2010.

Maldarelli, Claire. "Although Purebred Dogs Can Be Best in Show, Are They Worst in Health?" *Scientific American*, February 2014.

Manning, Sue. "Kittens Dropped at Shelters Are Often Euthanized." Associated Press, April 9, 2014.

Marek, Ramona. "Experts Find New Ways to Assess Pain." *Catnip*, September 2016.

Marek, Ramona. "Purring: The Feline Mystique." *Catnip*, September 2015.

Martin, P., and P. Bateson. "The Ontogeny of Locomotor Play Behaviour in the Domestic Cat." *Animal Behaviour* 33, 1985.

Mayon-White, Richard. "Pets—Pleasures and Problems." *The BMJ* (formerly *British Medical Journal*) 331, 2005.

McCleery, Robert A., et al. "Understanding and Improving Attitudinal Research in Wildlife Sciences." *Wildlife Society Bulletin* 34, 2006.

McComb, Karen, et al. "The Cry Embedded within the Purr." *Current Biology* 19, 2009.

McDonald, Jennifer L., et al. "Reconciling Actual and Perceived Rates of Predation by Domestic Cats." *Ecology and Evolution* 5, 2015.

McNicholas, June, et al. "Pet Ownership and Human Health: A Brief Review of Evidence and Issues." *The BMJ* (formerly *British Medical Journal*) 331, 2005.

Medina, Félix M., et al. "A Global Review of the Impacts of Invasive Cats on Island Endangered Vertebrates." *Global Change Biology* 17, 2011.

Meehl, Cindy, dir. *Buck* (documentary film). Georgetown, CT: Cedar Creek Productions LLC, 2011.

Mendl, Michael, and Robert Harcourt. "Individuality in the Domestic Cat." In Turner and Bateson, 1988.

Merola, Isabella, and Daniel S. Mills. "Behavioural Signs of Pain in Cats: An Expert Consensus." *PLOS ONE* 11, 2016.

Mertens, C., and D. C. Turner. "Experimental Analysis of Human–Cat Interactions during First Encounters." *Anthrozoös* 2, 1988.

Mertens, Claudia. "Human–Cat Interactions in the Home Setting." *Anthrozoös* 4, 1991.

Milius, Susan. "Social Cats." *Science News* 160, 2001.

Miller, Mary Anne. "How to Find a Lost Cat." *MyPetMD.com*, n.d.

Miller, Paul. "Why Do Cats Have an Inner Eyelid as Well as Outer Ones?" *Scientific American*, November 20, 2006.

Miller, Philip S., et al. "Simulating Free-Roaming Cat Population Management Options in Open Demographic Environments." *PLOS ONE* 10, 2014.

Moelk, Mildred. "The Development of Friendly Approach Behavior in the Cat: A Study of Kitten–Mother Relation and the Cognitive Development of the Kitten from Birth to Eight Weeks." *Advances in the Study of Behavior* 10, 1979.

Moelk, Mildred. "Vocalizing in the House Cat; A Phonetic and Functional Study." *American Journal of Psychology* 57, 1944.

Monk, Caroline. "The Effects of Group Housing on the Behavior of Domestic Cats (*Felis silvestris catus*) in an Animal Shelter." Honors Thesis. Ithaca, NY: College of Agricultural and Life Sciences, Animal Science, Cornell University, 2008.

Montague, Michael J., et al. "Comparative Analysis of the Domestic Cat Genome Reveals Genetic Signatures Underlying Feline Biology and Domestication." *Proceedings of the National Academy of Sciences* 111, December 2, 2014.

Montana Pioneer. "Livingston Man Sentenced: Durfey Case Generates National Interest." *Montana Pioneer,* January 2007.

Mooney, Chris. "The Science of Why We Don't Believe Science." *Mother Jones,* May/June 2011.

Morell, Virginia. *Animal Wise: The Thoughts and Emotions of Our Fellow Creatures.* New York: Crown, 2013.

Morris, James G., and Quinton R. Rogers. "Assessment of the Nutritional Adequacy of Pet Foods through the Life Cycle." *Journal of Nutrition*, December 1994.

Morris, James G. "Idiosyncratic Nutrient Requirements of Cats Appear to Be Diet-Induced Evolutionary Adaptations." *Nutrition Research Reviews* 15, 2002.

Mott, Maryann. "U.S. Faces Growing Feral Cat Problem." *National Geographic News,* October 28, 2010.

Mishra, Vyvyan, with Bonnie Schroeder. *Measuring the Benefits: Companion Animals and the Health of Older Persons.* Toronto: International Federation on Aging, 2014.

Nagelschneider, Mieshelle. *The Cat Whisperer: Why Cats Do What They Do—and How to Get Them to Do What You Want.* New York: Bantam, 2013.

Natoli, Eugenia, Alessandra Baggio, and Dominique Pontier. "Male and Female Agonistic and Affiliative Relationships in a Social Group of Farm Cats *(Felis catus* L.)." *Behavioural Processes* 53, 2001.

Natoli, Eugenia, and Emanuele De Vito. "The Mating System of Feral Cats Living in a Group." In Turner and Bateson, 1988.

Natoli, Eugenia, Emanuele De Vito, and Dominique Pontier. "Mate Choice in the Domestic Cat *(Felis silvestris catus* L.)." *Aggressive Behavior* 26, 2000.

Natoli, Eugenia, et al. "Management of Feral Domestic Cats in the Urban Environment of Rome (Italy)." *Preventive Veterinary Medicine* 30, 2006.

Natoli, Eugenia, et al. "Relationships between Cat Lovers and Feral Cats in Rome." *Anthrozoös* 12, 1999.

Natoli, Eugenia. "Urban Feral Cats *(Felis catus* L.): Perspectives for a Demographic Control Respecting the Psychobiological Welfare of the Species." *Annali dell'Istituto Superiore di Sanità* 30, no. 2, 1994.

Newman, Aline Alexander, and Gary Weitzman. *How to Speak Cat: A Guide to Decoding Cat Language.* Washington, DC: National Geographic Society, 2015.

Nogales, Manuel, et al. "A Review of Feral Cat Eradication on Islands." *Conservation Biology* 18, April 2004.

Nogales, Manuel, et al. "Feral Cats and Biodiversity Conservation: The Urgent Prioritization of Island Management." *BioScience* 63, 2013.

Nutter, Felicia B., Jay F. Levine, and Michael K. Stoskopf. "Reproductive Capacity of Free-Roaming Domestic Cats and Kitten Survival Rate." *Journal of the American Veterinary Medical Association* 225, 2004.

O'Brian, Patrick. *The Far Side of the World.* London: William Collins Sons & Co. Ltd., 1984.

Owren, Michael J., Drew Rendall, and Michael J. Ryan. "Redefining Animal Signaling: Influence versus Information in Communication." *Biological Philosophy* 25, 2010.

Patronek, Gary J., A. M. Beck, and L. T. Glickman. "Dynamics of Dog and Cat Populations in a Community." *Journal of the American Veterinary Medical Association,* March 1, 1997.

Patronek, Gary J. "Mapping and Measuring Disparities in Welfare for Cats across Neighborhoods in a Large U.S. City." *American Journal of Veterinary Research* 71, February 2010.

Paul, Caroline. *Lost Cat: A True Story of Love, Desperation, and GPS Technology.* New York: Bloomsbury, 2013.

Paxton, Tom. "The Last Thing on My Mind" (lyrics). United Artists Music Co., 1964.

People for the Ethical Treatment of Animals. "Animal Rights Uncompromised: Feral Cats." *Peta.org,* n.d.

People for the Ethical Treatment of Animals. "Feral Cats: Trapping Is the Kindest Solution." *Peta.org,* n.d.

Peterson, Ernest A., W. Carlos Heaton, and Sydney W. Wruble. "Levels of Auditory Response in Fissiped Carnivores." *Journal of Mammalogy* 50, 1969.

Peterson, M. Nils, et al. "Opinions from the Front Lines of Cat Colony Management Conflict." *PLOS ONE* 7, 2012.

PetMD.com. "How to Walk a Cat (and Live to Tell about It)." www.petmd.com/cat/training/evr_ct_how_to_walk_you_walk_your_cat?page=show, n.d.

Pontier, Dominique, et al. "Retroviruses and Sexual Size Dimorphism in Domestic Cats." *Proceedings of the Royal Society B* 265, 1998.

Potter, Alice, and Daniel Simon Mills. "Domestic Cats (*Felis silvestris catus*) Do Not Show Signs of Secure Attachment to their Owners." *PLOS ONE* 10, 2015.

Poucet, B. "Spatial Behavior of Cats in Cue-Controlled Environments." *Quarterly Journal of Experimental Psychology* 37, 1985.

Provincia di Roma. "Il decalogo della perfetta gattara" in *Mici Amici: Una Guida ai Doveri, agli Obblighi, ma Anche ai Diritti per una Convivenza Solidale e Informata fra Gatti e Umani.* Rome: Province of Rome, Office of Animal Protection, 2006.

Pryor, Karen. *Getting Started: Clicker Training for Cats.* Waltham, MA: Karen Pryor Books, 2003

Quackenbush, James, and Lawrence Glickman. "Social Work Services for Bereaved Pet Owners: A Retrospective Case Study in a Veterinary Teaching Hospital" in Katcher and Beck, 1983.

Quackenbush, Roger E. "Genetics of the Domestic Cat: A Lab Exercise." *The American Biology Teacher* 54, 1992.

Quammen, David. *The Flight of the Iguana.* New York: Delacorte Press, 1988.

Quito, Anne. "Glass Architecture Is Killing Millions of Migratory Birds." *Citylab.com*, March 31, 2015.

Qureshi, Adnan, et al. "Cat Ownership and the Risk of Fatal Cardiovascular Diseases." *Journal of Vascular and Interventional Neurology*, January 2009.

Ramzy, Austin. "Australia Defends Plan to Cull Cats." *New York Times*, October 16, 2015.

Ratcliffe, Norman, et al. "The Eradication of Feral Cats from Ascension Island and Its Subsequent Recolonization by Seabirds." *Oryx* 44, 2009.

Reis, Pedro M., et al. "How Cats Lap: Water Uptake by *Felis catus*." *Science*, November 26, 2010.

Remitz, Jessica. "Defining Senior Age in Cats." *PetMD.com*, April 21, 2016.

Remitz, Jessica. "Does My Senior Cat Hate Me?" *Pet360.com*, March 8, 2016.

Rochlitz, Irene. "A Review of the Housing Requirements of Domestic Cats (*Felis silvestris catus*) Kept in the Home." *Applied Animal Behaviour Science*, September 2005.

Roe, Barbara. "Keeping a Bobcat or Canadian Lynx as a Pet." Bitterroot Bobcat & Lynx, Stevensville, Montana, n.d.

Rollin, Bernard E. "Morality and the Human–Animal Bond." In Katcher and Beck, 1983.

Royal Society for the Protection of Birds. "Are Cats Causing Bird Declines?" https://wwwrspb.org.uk/get-involved/community-and-advice/garden-advice/unwantedvisitors/cats/birddeclines.aspx.

Savage, R. J. G., and M. R. Long. *Mammal Evolution—An Illustrated Guide.* New York: Facts on File, 1986.

Say, Ludovic, Dominique Portier, and Eugenia Natoli. "High Variation in Multiple Paternity of Domestic Cats (*Felis catus* L.) in Relation to Environmental Conditions." *Proceedings of the Royal Society B*, 1999.

Schmidt, Paige M., et al. "Evaluation of Euthanasia and Trap-Neuter-Return (TNR) Programs in Managing Free-Roaming Cat Populations." *Wildlife Research* 36, 2009.

Schötz, Susanne, and Joost van de Weijer. "Human Perception of Intonation in Domestic Cat Meows." *Proceedings from Fonetik 2014*, Department of Linguistics, Stockholm University.

Schötz, Susanne, and Robert Eklund. "A Comparative Acoustic Analysis of Purring in Four Cats." *Proceedings from Fonetik 2011*, Department of Speech, Music, and Hearing, and Centre for Speech Technology, Royal Institute of Technology, Stockholm.

Schötz, Susanne. "A Phonetic Pilot Study of Chirp, Chatter, Tweet, and Tweedle in Three Domestic Cats." *Proceedings of Fonetik 2013*, Linköping University, Sweden.

Schötz, Susanne. "A Phonetic Pilot Study of Vocalisations in Three Cats." *Fonetik 2012: Proceedings of the XXVth Swedish Phonetics Conference,* May 30–June 1, 2012.

Segelken, Roger. "It's the Cat's Meow: Not Language, Strictly Speaking, but Close Enough to Skillfully Manage Humans, Communication Study Shows." *Cornell Chronicle,* May 20, 2002.

Seymour, Gene. "They're Purr-fect as Pals: Temple Research Project Matching Pussycats with People." Philadelphia *Daily News,* March 5, 1986.

Shepherdson, D. J., et al. "The Influence of Food Presentation on the Behavior of Small Cats in Confined Environments." *Zoo Biology* 12, 1993.

Siegal, Mordecai, ed. *The Cornell Book of Cats: A Comprehensive Medical Reference for Every Cat and Kitten.* New York: Villard Books, 1991.

Silva-Rodriguez, Eduardo A., and Kathryn E. Sieving. "Influence of Care of Domestic Carnivores on their Predation on Vertebrates." *Conservation Biology* 25, 2011.

Singer, Jo. "How to Recognize an Emergency." *Catnip* 24, March 2016.

Skoglund, Pontus, et al. "Ancient Wolf Genome Reveals an Early Divergence of Domestic Dog Ancestors and Admixture into High-Latitude Breeds." *Current Biology* 25, 2015.

Slack, Gordy. "The Rights (and Wrongs) of Cats." *California Wild, the Magazine of the California Academy of Sciences* 51, 1998.

Spinka, Marek, Ruth C. Newberry, and Marc Bekoff. "Mammalian Play: Training for the Unexpected." *The Quarterly Review of Biology* 76, 2001.

Squires, Nick. "Stray Cat Colony in Ancient Roman Temple Is Declared a Health Hazard." *The Telegraph,* November 3, 2012.

Stewart, Mary. "Loss of a Pet—Loss of a Person." In Katcher and Beck, 1983.

Sugden, Karen, et al. "Is Toxoplasma Gondii Infection Related to Brain and Behavior Impairments in Humans? Evidence from a Population-representative Birth Cohort." *PLOS ONE* 11, February 2016.

Sydney Morning Herald. "Savannah Cats Banned from Australia," August 3, 2008.

Taylor, R. H. "How the Macquarie Island Parakeet Became Extinct." *New Zealand Journal of Ecology* 2, 1979.

Tellington-Jones, Linda. *The Tellington TTouch for Happier, Healthier Cats, DVD. Ttouch.com.*

Tellington-Jones, Linda, with Sybil Taylor. *The Tellington TTtouch: Caring for Animals with Heart and Hands.* New York: Penguin Books, 1992.

Thomas, Ben. "What's So Special about Mirror Neurons?" *blogs .scientificamerican.com,* November 6, 2012.

Thomas, Elizabeth Marshall. *The Tribe of Tiger: Cats and Their Culture.* New York: Simon and Schuster, 1994.

Thompson, Andrea. "What's the Most Popular Pet?" *Livescience.com,* January 15, 2013.

Thornton, Alex, and Dieter Lukas. "Individual Variation in Cognitive Performance: Developmental and Evolutionary Perspectives." *Philosophical Transactions of the Royal Society B* 367, 2012.

Tolford, Katherine. "Do Cats Dream?" *Pet360.com.* July 26, 2016.

Toukhatsi, Samia R., Pauleen C. Bennett, and Grahame J. Coleman. "Behaviors and Attitudes toward Semi-Owned Cats." *Anthrozoös* 20, 2007.

Trumps, Valerie. "Fifteen Surefire Ways to Bond with Your Cat." *Pet360.com,* August 2015.

Tschanz, Britta, et al. "Hunters and Nonhunters: Skewed Predation Rate by Domestic Cats in a Rural Village." *European Journal of Wildlife Research* 57, June 2011.

Tucker, Arthur O., and Sharon S. Tucker. "Catnip and the Catnip Response." *Economic Botany* 42, 1988.

Turner, Dennis C., and Gerulf Rieger. "Singly Living People and their Cats, a Study of Human Mood and Subsequent Behavior." *Anthrozoös* 14, 2001.

Turner, Dennis C., and Othmar Meister. "Hunting Behaviour of the Domestic Cat." In Turner and Bateson, 1988.

Turner, Dennis C., and P. P. G. Bateson. *The Domestic Cat: the Biology of Its Behaviour.* Cambridge, UK: Cambridge University Press, 1988.

Turner, Dennis C., Gerulf Rieger, and Lorenz Gygax. "Spouses and Cats and their Effects on Human Mood." *Anthrozoös* 16, 2003.

Turner, Dennis C. "Noncommunicable Diseases: How Can Companion Animals Help in Connection with Coronary Heart Disease, Obesity, Diabetes, and Depression?" In J. Zinsstag et al., *One Health: The Theory and Practice of Integrated Health Approaches.* Oxfordshire, UK: CAB International, 2015.

Turner, Dennis. C. "The Ethology of the Human–Cat Relationship." *Schweizer Archiv für Tierheilkunde* 133, 1991.

University of Illinois at Champaign-Urbana. "Cats Pass Diseases to Wildlife, Even in Remote Areas." *Science Daily*, May 12, 2011.

Van Neer, Wim, Veerle Linseele, Renée Friedman, and Bea De Cupere. "More Evidence for Cat Taming at the Predynastic Elite Cemetery of Hierakonpolis (Upper Egypt)." *Journal of Archaeological Science* 45, 2014.

Vigne, Jeyan-Denis, et al. "Earliest 'Domestic' Cats in China Identified as Leopard Cat (*Prionailurus bengalensis*)." *PLOS ONE* 11, January 22, 2016.

Villalobos, Alice. "Quality of Life Scale" from *Canine and Feline Geriatric Oncology: Honoring the Human–Animal Bond.* Ames, IA: Blackwell Publishing, 2007. Revised for the International Veterinary Association of Pain Management 2011 Palliative Care and Hospice Guidelines.

Vogelsang, Jessica. "How to Discover Your Pet's Secret Pain." *PetMD .com*, December 31, 2015.

Vogelsang, Jessica. " 'Miracle' Technology Is Available to Save Your Pet, but Can You Afford It?" *PetMD.com*, February 19, 2016.

Vogt, Amy Hoyumpa, et al. "Feline Life Stage Guidelines." *Journal of Feline Medicine and Surgery* 12, 2010.

Voith, Victoria L., and Peter L. Borchelt. "Social Behavior of Domestic Cats." In Voith, V. L., and P. L. Borchelt, eds., *Readings in Companion Animal Behavior.* Trenton, NJ: Veterinary Learning Systems, 1996.

Voith, Victoria L. "Attachment of People to Companion Animals." *Veterinary Clinics of North America: Small Animal Practice* 15, no. 2, 1985.

Wade, Nicholas. "DNA Offers New Insight Concerning Cat Evolution." *New York Times,* January 6, 2006.

Walker, James. "Wildcat Haven—Saving a Species on the Brink of Extinction." *Conservation-careers.com,* December 31, 2015.

Weir, Kirsten. "The Cat Whisperer." *Salon.com,* March 18, 2008.

Whilde, A. "The Prey of Two Rural Domestic Cats." *The Irish Naturalists' Journal* 24, 1992.

Whitaker, Julie, and William Steinkraus: *The Horse: A Miscellany of Equine Knowledge.* Lewes, East Sussex, UK: Ivy Press Ltd., 2007.

Wilken, Rachel L. M. "Feral Cat Management: Perceptions and Preferences (A Case Study)." M.S. Thesis. San Jose State University, 2012.

Willson, S. K., I. A. Okunlola, and J. A. Novak. "Birds Be Safe: Can a Novel Cat Collar Reduce Avian Mortality by Domestic Cats (*Felis catus*)?" *Global Ecology and Conservation,* January 2015.

Wiman, Christian. *My Bright Abyss: Meditation of a Modern Believer.* New York: Farrar, Straus and Giroux, 2013.

Yates, Diana. "Researchers Track the Secret Lives of Feral and Free-Roaming House Cats." University of Illinois News Bureau, 2011.

Zak, Paul. "Dogs (and Cats) Can Love." *The Atlantic,* April 2014.

Index

Index

Index

Index